Canadian Wings

edited by

STEPHEN PAYNE, CURATOR

CANADA AVIATION MUSEUM

featuring the artwork of

ROBERT W. BRADFORD, CM

& DAN PATTERSON

Canadian Wings

A REMARKABLE CENTURY OF FLIGHT

Canada Aviation Museum

Douglas & McIntyre

VANCOUVER/TORONTO/BERKELEY

06 07 08 09 10 5 4 3 2 1

Douglas & McIntyre Ltd.
2323 Quebec Street, Suite 201
Vancouver, British Columbia
Canada V5T 4S7
www.douglas-mcintyre.com

Cet ouvrage a été publié simultanément en français sous le titre
Les ailes du Canada: Un siècle d'aviation prodigieux

Library and Archives Canada Cataloguing in Publication
Canadian wings : a remarkable century of flight /
Stephen Payne, editor, Canada Aviation Museum ;
featuring the artwork of Robert W. Bradford and Dan Patterson.

Co-published by the Canada Aviation Museum.
Includes bibliographical references and index.
ISBN-13: 978-1-55365-167-3 · ISBN-10: 1-55365-167-7

1. Aeronautics—Canada—History—20th century. I. Bradford, R. W.
II. Patterson, Dan, 1953– III. Payne, Stephen R. IV. Canada Aviation Museum
TL523.C365 2006 629.13′0971′0904 C2006-900359-9

Editing by Jonathan Dore
Jacket and text design by Peter Cocking
Jacket photographs: Front: Details from *Critical Approach* by Charles Vinh, 1992 (top) and A.E.A. *Silver Dart* by Robert Bradford, CM, 1965 (bottom). Back: Details from *Canadian Vickers Vedette Mk. Va over Montreal*, 1963 (top) and *Bob Cockram's Noorduyn Norseman Mk. II: Ruth IV*, 1963–64 (bottom) by Robert Bradford, CM.
Frontispiece: Detail from front cover, *Star Weekly*, January 6, 1940;
photographed by Gerald Richardson, staff photographer.
Printed and bound in China by C&C Offset Printing Co., Ltd.
Printed on acid-free paper
Distributed in the U.S. by Publishers Group West

We gratefully acknowledge the financial support of the Canada Council for the Arts, the British Columbia Arts Council, and the Government of Canada through the Book Publishing Industry Development Program (BPIDP) for our publishing activities.

Contents

Foreword

AVIATION *in* CANADA'S CENTURY

JUST OVER a century ago, Prime Minister Sir Wilfrid Laurier confidently predicted that the twentieth century belonged to Canada. Although academics and political commentators have often challenged this claim, it is without doubt that the development of Canada over much of the last hundred years has been due to the contributions made by aviation.

The centennial of powered flight in Canada is an opportunity for Canadians to celebrate all that has been accomplished in the field of aviation over a century. From that first short flight from the ice of Bras d'Or Lake in Nova Scotia by the *Silver Dart* on February 23, 1909, to the present day, aviation has grown from Alexander Graham Bell's small group of adventurous experimenters to a massive industry. It has served to explore, develop, and defend this enormous country, tying it together and linking it with the world. Today, Canada stands high among the leading countries of the world in the quality and quantity of its aviation activities, in fields such as general aviation, commercial air transport, and airport and navigation infrastructure, as well as in the manufacture of aircraft, engines, components, and simulators.

For over forty years, the Canada Aviation Museum and its predecessor, the National Aviation Museum, have been collecting, preserving, displaying, and interpreting aircraft and other artifacts that represent the important milestones in our aviation history. From its first exhibit hall at Ottawa's Uplands air terminal in 1960 through its move to the historic Rockcliffe airfield in 1964 and finally the opening of its massive exhibit building in 1988, the Museum has continued to grow, both in the size of its collection and in its capacity to present it to the public.

In the years that have followed, the Museum has added many important items to its collection. Perhaps the most significant new aircraft acquisition was the 1912 Borel Morane monoplane, the oldest existing aircraft to have flown in Canada. More modern acquisitions include a German Messerschmitt BF 109 fighter, a Douglas DC-9 airliner, a Boeing Vertol CH-113 Labrador search-and-rescue helicopter, and an early example of the Canadian Armed Forces current supersonic fighter aircraft, the McDonnell Douglas CF-18. The Museum has also vastly increased the scope of its activities in areas such as exhibitions, educational programs, and special public events, and has created Canada's foremost website in its field (www.aviation.technomuses.ca).

The continuing challenge of inadequate storage space for a growing aircraft collection was evident even as the new Museum building opened its doors in 1988, when several larger aircraft had to be left parked outside. The situation was finally rectified early in 2005, when a new 8 000-square-metre (86,111-square-foot) storage building was opened along with new administrative, library, and archives facilities. The Museum has now redeveloped the part of the main building formerly used for storage to house new exhibits and provide additional space for public programs.

While the Museum has made enormous progress in the preservation and presentation of our aviation heritage, the contribution that aviation has made to the development of Canada during the twentieth century is still not as well known to Canadians as it should be. Accordingly, the Museum is creating a comprehensive exhibit and website on Canadian aviation accomplishments for the 2009 centennial. We hope this book, too, will serve to enhance Canadians' aviation knowledge and appreciation of their aviation history.

ANTHONY P. SMYTH
Director General · Canada Aviation Museum

Acknowledgements

THE CREATION OF a book such as this naturally involves the input of a great number of people. The Museum is grateful to everyone who participated and helped to bring this book to fruition.

First, we would like to thank Hugh Halliday for his enormous contribution of researching and writing early drafts of the book. Chapter 11 on aircraft manufacturing was based on the great body of research conducted by the late Fred Shortt. The Museum's team of Anthony Smyth, Renald Fortier, Fiona Smith Hale, and Wendy McPeake oversaw the long march to publication. Andrew MacDonald cheerfully undertook the arduous task of scanning all the photographs and paintings in the book. We are thankful to Didier Feminier for his graceful translation, Elizabeth Macfie and Louise Saint-André for their work on the index, and Lyne Tardif for proofreading the French version.

We are especially indebted to Robert Bradford, CM, for his many works of art included in the book as well as to Charles Vinh, Frank Oord, James Leech, and Helene Croft for permission to reproduce their works in the Museum's collection. We are also indebted to noted aviation photographer Dan Patterson for his creative portrayal of Museum artifacts in the book. Jeff Leavens and Ray Cavan helpfully provided other needed photos as did Richard Dawe at Cougar Helicopters and Lorraine Ste-Marie at Bombardier. Janet Lacroix at the Canadian Forces Joint Imaging Centre was very helpful in making their central negative library accessible for research. Justin Cuffe at Canada's Aviation Hall of Fame gave generously of his time to provide some rare photos as did the staffs of Library and Archives Canada and the Canadian War Museum, making images available from their collections.

Finally, at Douglas & McIntyre, Jonathan Dore brought his thoughtful editing skills to the project; Susan Rana kept us all on schedule; Peter Cocking gave us the beautiful design and layout, and Naomi Pauls of Paper Trail Publishing did a fine job proofreading. We thank Scott McIntyre for taking on this project and encouraging us to the end.

STEPHEN PAYNE
Curator, Aeronautical Technology · Canada Aviation Museum

FTBJ

Introduction

a MUSEUM *in the* MAKING

AVIATION HAS A very long history, extending back before the first heavier-than-air flights to ballooning and even to the myths that expressed humanity's earliest dreams of flight and its awe at the abilities of birds. Canada's participation dates from 1840 and includes the creation of a national museum dedicated to preserving Canada's aeronautical heritage. Now located at the historical site of Rockcliffe in Ottawa, the Canada Aviation Museum is a rich repository of all things aeronautical and Canadian.

facing page: A Museum staff member puts protective coverings on the aircraft shortly before the Canada Aviation Museum's 1988 opening.

ROCKCLIFFE: A SITE FOR THE MUSEUM

The airfield at the former Canadian Forces Base Rockcliffe, now the Museum's address, was initially farmland. In 1895, the Dominion government began expropriating the property below the limestone cliffs for a militia rifle range. Ten years later, the area above the cliffs was acquired as well. The nearby village of Rockcliffe lent its name to the site, which became the local centre for shooting competitions and militia exercises. On the outbreak of the First World War in 1914, Rockcliffe was used as an encampment for at least two battalions recruited and drilled for overseas service. The 1st Motor Machine Gun Brigade, which pioneered Canadian armoured cars, was also organized at the base in September 1914. In August and September 1918, the Royal Air Force carried out three round-trip airmail flights between Toronto and Ottawa using Curtiss JN-4 (Can.) training aircraft and *facing page, top:* Aviation activity at Rockcliffe began with the first experimental airmail flights between Ottawa and Toronto in August 1918. Third from the right is William Lyon Mackenzie King, the future prime minister.

facing page, bottom: A de Havilland DH.4 of the Canadian Air Board about to leave on an airmail flight from Rockcliffe to Toronto in July 1923. The earlier temporary canvas hangar has given way to a more permanent building.

the Rockcliffe Rifle Ranges as their Ottawa airfield, so it is fitting that the Canada Aviation Museum now displays a JN-4.

Following the First World War, Rockcliffe became the focal point for aviation in the capital. With flat land for wheeled aircraft, the nearby river for seaplanes, and a location safely out in the countryside away from the city, it was ideal for aircraft. The first Canadian experiments in aerial photography were conducted from Rockcliffe in the summer of 1920. On the ground, temporary tube-and-canvas structures gave way to more permanent buildings. From 1920 onward, the Canadian Air Force—by one name or another—was in continuous occupation of the site. By 1927, the last of the militia rifle butts had been removed. Thereafter, until a Royal Canadian Mounted Police (RCMP) Air Division was formed in 1937, Rockcliffe was an all–Royal Canadian Air Force (RCAF) establishment.

The work carried out was as varied as there were uses for aircraft. Pilots and scientists struggled with problems that ranged from experimental crop dusting to developing special lubricants for winter flying. Above all, Rockcliffe became the centre for aerial mapping in Canada. Every summer, photographic detachments fanned out across the nation, flying along predetermined lines, photographing the terrain, and sending the films back to Ottawa to be developed for conversion into charts. In recognition of this important work a specialized Photographic Establishment was opened in 1936.

During the 1930s, the site was enlarged and developed, much of the work being done as relief projects for the unemployed. When the Second World War broke out in 1939, Rockcliffe continued to serve as an experimental flight test centre, as a Manning Depot for new recruits, and as the primary base for RCAF mail flights overseas.

Aerial mapping was curtailed during the war but resumed in the summer of 1945, continuing apace until the 1960s. Much of this work was performed by Lancaster bombers converted to the photo-survey role, a classic case of swords beaten into ploughshares. Flight tests continued until 1957, when the Central Experimental and Proving Establishment moved across the city to Uplands, with its longer runways. A Piston-Engine Practice Flight continued to operate from Rockcliffe until 1966—the last RCAF unit to use the airfield. Between July 1974 and April 1976, the airfield was used by the government's trial Airtransit service to operate short-takeoff-and-landing (STOL) flights to Montréal;

FIRST AERIAL MAIL SERVICE
PER
ROYAL AIR FORCE
OTTAWA TO TORONTO
AUG. 27. 1918. KA-308

it was to be the airport's only commercial application. Airtransit offices were on the site now occupied by the Canada Aviation Museum. To this day, the Rockcliffe Flying Club keeps on-site aerial activity alive, and many of the Museum's donated aircraft first arrive by flying in to the airport.

Rockcliffe, then, is a rich source of Canada's flying heritage, with its history of militia training, flight testing, air transport, and aerial photography, once a STOL airport and now the site of the Canada Aviation Museum.

That a museum exists at all is due to the vision and efforts of several key individuals. The first was Sir Arthur Doughty (1860–1936), who, as both Dominion archivist and an officer in the Canadian Expeditionary Force, acquired a large number of "war trophies" after the First World War. These included the Museum's examples of the Royal Aircraft Factory BE.2C, Junkers J.I, AEG G.IV, and the remains of William Barker's Sopwith Snipe. Ultimately, Doughty's efforts would benefit both the Canadian War Museum (organized in 1942) and the future Canada Aviation Museum.

John H. Parkin (1891–1981) had been keenly interested in aviation as early as 1911. In 1929, he was appointed to the National Research Council (NRC) as assistant director of the Division of Physics, with responsibilities for aeronautical research. Parkin encouraged the collection and preservation of artifacts for a future museum. He had several supporters, notably Group Captain (later Air Vice-Marshal) Ernest W. Stedman (1888–1957), the RCAF's first chief aeronautical engineer, and Major-General Andrew McNaughton (1887–1966), chief of the General Staff and, from 1935 to 1939, president of the NRC. The group continued to acquire smaller artifacts, but storage space was very limited; the sole complete aircraft accessioned during this period was a Sopwith Camel. The only display space available was a room at NRC headquarters on Sussex Drive. Even that facility was closed during the Second World War.

Following the war, the RCAF set aside a number of aircraft for a museum. After three years of discussions, a museum committee was finally struck in 1953, with representatives from the National Research Council, the Defence Research Board, the Department of Transport, the Department of National Defence, the Royal Canadian Flying Clubs Association, the Air Cadet League of Canada, and the Department of Northern Affairs and Natural Resources.

But a committee was not a museum. Two events and one man were crucial to one being formed. The events were the fiftieth anniversary of powered flight in Canada in 1959 and the construction of a new air terminal at Uplands with space available for a museum. Malcolm S. "Mac" Kuhring, who headed the NRC's Engine Laboratory, drafted the first plans and pushed all the right political buttons. On October 25, 1960, the National Aviation Museum (as it was then called) was formally opened in the Uplands terminal.

The first curator of the new Museum was Kenneth M. Molson (1916–1996). As curator of the National Aviation Museum, Molson expanded the collection until his retirement in 1967. He was a prolific writer of books and magazine articles, and his photographic collection and personal papers, assembled over decades, are now part of the Museum's archives. Wing Commander Ralph Manning (1916–1994), the RCAF's official air historian, also took great interest in the museum and arranged for many stored aircraft to be brought to eastern Canada. Molson and Manning then recruited Lee Murray, director of the Canadian War Museum, and the resources of the two museums and the RCAF were pooled to create the National Aeronautical Collection in the mid-1960s.

The establishment of the National Aeronautical Collection coincided with the RCAF abandoning air operations at Rockcliffe. Hangars that had previously housed Avro Lancasters and Beechcraft Expeditors were now available for permanent storage and display. In 1968, formal custody of the aircraft was transferred to the Aviation

facing page: By the 1930s, Rockcliffe had become the centre for aerial mapping in Canada. Here, a Fairchild camera is mounted in the nose of a Vickers Vedette in 1931.

below: The Aerial Experiment Association *Silver Dart* replica on display in the Museum's exhibit area. This aircraft flew at Baddeck, Nova Scotia, in 1959 to commemorate the fiftieth anniversary of the first powered flight in Canada by its namesake predecessor.

facing page: The first phase of the Museum's expansion included the construction of the present exhibition building, centre left, seen in this 1989 view. The Second World War–era hangars, which previously served as the collection's home, are being demolished in the background.

and Space Division of the newly established National Museum of Science and Technology. The Museum's original name was restored in 1982, then replaced by its current title—the Canada Aviation Museum—in 2002. Meanwhile, in 1972, the National Air Museum Society was formed (and still exists) to encourage public support for the Museum and its objectives.

The move of the National Aeronautical Collection to Rockcliffe allowed partial consolidation of the aircraft that had hitherto been scattered across several sites. However, the hangars were obvious firetraps, and National Aviation Museum staff were relieved when, in June 1988, they took possession of phase one of a new, purpose-built museum. With the aircraft collection housed in secure quarters, the way was clear for the Museum to pursue its ambitious plan to present Canada's rich aviation history to the world.

It also offered an attractive venue for special events such as the anniversary of the Battle of Britain, Canada Day, and visits from "celebrity" aircraft. Band concerts, aerobatic displays by the Snowbirds, tethered hot-air balloon ascents, parachute-team demonstrations, miniature air shows featuring radio-controlled model aircraft, and passenger flights in de Havilland Canada Chipmunk and Stearman Model 75 aircraft have been regular features of the Museum's public program.

THE COLLECTION

The Canada Aviation Museum is recognized as having the most extensive aviation collection in Canada, one that ranks among the best in the world. Original aircraft reflect the evolution of aeronautical science with special reference to Canada. Although the aircraft are the highlight, the Museum also maintains an extensive collection of models, aviation art, photographs, films, engines, propellers, instruments, flying gear, memorabilia, and library and archival materials. Over the years, this collection has been built through donation, purchase, transfer, and exchange.

The first acquisitions were the few surviving First World War "trophies." Neither the RCAF nor the NRC attempted to preserve any aircraft of the 1919–39 era, but following the Second World War, the RCAF stored several retired aircraft, including fighters such as a Hawker Hurricane and training machines such as a rare Fairey Battle with gun turret. Captured German machines brought to Canada include two Heinkel HE 162 jets and two Messerschmitt ME 163 Komet rocket aircraft.

Between 1918 and 1955 many examples of historic aircraft were lost or destroyed through oversight or lack of storage space. Many of the resulting deficiencies have subsequently been made good through the acquisition of replica aircraft manufactured from scratch. The *Silver Dart* replica is just such an example, its builders taking inspiration from a non-flying replica assembled at Trenton in 1958. The RCAF resolved to build a flying replica for the fiftieth anniversary of the first Canadian powered flight the following year; two new replicas were in fact assembled, one for static display and one for flying. The latter was shipped to Baddeck, Nova Scotia, where a re-enactment of J.A.D.

Karl the Flying Trapper, by Helene Croft. This painting is housed in the Museum's art collection, which has continued to grow through donation and purchase.

McCurdy's historic flight took place on February 23, 1959, with McCurdy himself present. The *Silver Dart* replica was very unstable and crashed in gusty winds. Happily, the pilot, Wing Commander Paul Hartman, was not hurt.

The *Silver Dart* was followed by other replicas, most notably the Nieuport 17, Sopwith Pup, and Sopwith Triplane. These were flown at various air displays until it was decided that the risk of damage to the valuable artifacts was too great. A particularly impressive replica is the Museum's Curtiss HS-2L. It incorporates parts from three original HS-2LS (including Canada's first bush plane, G-CAAC) but is essentially an aircraft built "from the ground up."

The acquisition of every machine has a story, and some are remarkable indeed. The new National Aviation Museum had little hope of finding a Curtiss JN-4 (Can.) when inquiries began, but in May 1961 one was found stored in a New York barn. With scant funds available, the Museum paid $9,000 for the machine and a spare set of wings that had been left in the rafters.

New York State was just next door—Afghanistan was not. In 1974, the Museum set out to discover if any Hawker Hinds remained in that country, which had acquired several in the 1930s. As an example of advanced biplane technology, a Hind would be an interesting acquisition. The Canadian air attaché to Pakistan reported that he had found two Hawker Hind airframes and three engines in a Kabul junkyard. Museum officials flew to Pakistan and travelled through the Khyber Pass to Kabul, Afghanistan, where they inspected the aircraft. Afghan authorities assisted, offering a Hind as a gift to Canada. Dismantling and shipping the machine (and a second, incomplete airframe) later that year was an adventure in itself. Promised equipment proved unavailable or unserviceable, and civil officials were capricious about the export of the machines. The Hind was eventually transported to Kabul airport and flown to Ottawa aboard a Canadian Armed Forces Hercules, and it has since been beautifully restored

top: The Museum's Hawker Hind in Afghanistan, prior to being dismantled and shipped to Canada in 1975.

bottom: The Hind following completion of its four-year restoration to airworthiness in 1988. On many occasions, the engine has been ground-run for public demonstration.

Donations of another sort have added many valuable aircraft to the collection. A former fighter pilot, John Paterson, donated his Supermarine Spitfire IX to the Museum and offered to assist in further acquisitions. In 1965, a Stearman 4EM was offered for sale in Idaho, where it had been in use as a crop duster. At the Museum's request, Paterson bought the machine, restored it to a mailplane configuration, and delivered it to the Museum in 1970. In succeeding years, individuals, manufacturers, airlines, and the Canadian Forces have donated aircraft to the collection (see Appendix for a complete list of the Museum's aircraft).

Exchanges are rare, but at one time the Museum had two Messerschmitt ME 163B rocket fighters. One of these was used to acquire a Messerschmitt BF 109F from a British collector. In 1967, the Indian Air Force obtained a Westland Lysander through the auspices of the RCAF and the Museum and in return offered Canada a Consolidated *facing page:* The continuing challenge of inadequate storage space for a growing collection was rectified early in 2005, when a new 8 000-square-metre (86,111-square-foot) storage building was opened along with new administrative, library, and archives facilities.

Liberator, a type just then being retired from the Indian Air Force. The Liberator was flown in stages back to Canada with a Canadian Armed Forces crew in May and June 1968. The 16 000-kilometre (10,000-mile) trip was the longest ferry flight of any aircraft in the Museum's collection.

An aeronautical museum consists of more than its aircraft. The Canada Aviation Museum is no exception. In the course of building its collection of aircraft, the Museum began collecting books and archival materials to assist staff in interpreting and restoring its collection. Today, the Museum's library is the finest of its kind in Canada and is a busy research centre for aviation enthusiasts and historians from across Canada and around the world.

The library's holdings include over 10,000 books, 7,000 periodicals, and 7,500 technical manuals, as well as more than 35,000 research photographs and 200,000 negatives, all focussed on the history of civilian and military aeronautics, with emphasis on the Canadian experience. The library subscribes to 200 periodical titles and holds full runs of Canadian aviation journals, such as *Canadian Aviation* and the *Canadian Aviation Historical Society Journal,* as well as aviation journals from other countries. Many types of aircraft are also represented in the Museum's comprehensive collection of technical manuals, which provide unique information such as detailed operating procedures and maintenance instructions. The Austro-Hungarian Collection, purchased in 1964, is an important holding that provides superb coverage of aviation history prior to the First World War.

The large collection of mostly black-and-white photographs dates from the earliest days of heavier-than-air flight to the present day, the majority being from the 1920s to the 1940s and coming to the Museum from private and corporate collections.

The Museum has several corporate and private archival collections, providing a unique source for researchers. Companies such as Air Canada, Canadair, and Avro Canada are just a few of the many that have donated books, photographs, and manuals. Private collections include the logbooks of First World War and Second World War aviators, the correspondence and papers of bush pilot Stuart Graham, Ken Molson's collection, and many others.

The art collection includes works by Robert W. Bradford, CM (1923–), the Museum's second chief curator and later director (1967–89). His paintings were the basis of a pop-

ular series of prints, many of which are included in this book. In 2002, the Museum acquired drawings for magazine advertisements that he had executed while working for de Havilland Canada.

The art collection itself has continued to grow through donation and purchase. Since 1992, the Museum has hosted an annual competition entitled Artflight. Beginning with "Made in Canada" (forty-five works exhibited), each competition has adopted a broad theme. The event inspired the formation of the Canadian Aviation Artists' Association in 1996. The collection now includes works by artists James Bruce, Don Connolly, Robert Banks, Frank Wootton, Eric Aldwinckle, Patrick Cowley-Brown, and Kenneth Lochhead, demonstrating the diversity of the Museum's holdings.

The story of Canadian aviation is best told in a global context, and a visit to the Canada Aviation Museum's website (www.aviation.technomuses.ca) will demonstrate the breadth of the Museum's mandate—an international view of aviation history with a focus on Canada. The completion of its new storage and administration buildings in 2005, the first phase of an ongoing development project, provides improved viewing of the aircraft as well as an enlarged library and archives. Throughout its history, the Museum has maintained a constant goal—to preserve our aviation legacy for future generations and to acknowledge and celebrate remarkable aeronautical achievements in Canada and around the world.

BIG
CELEBRATION
FOR SOLDIERS AND OLD BOYS
MONDAY JUNE 21-22 TUESDAY
BARRIE
HIGHLANDERS' BAND
Dare-devil Landrigan
AVIATOR
MONTREAL MEET
JUNE 25–JULY 4th, 1910
Atlantic City
AERO CLUB
Aviation Meet
July 4th to 11th
1910
PRESS
INTERNATIONAL
AVIATION TOURNAMENT
Aviator
BELMONT PARK 1910
AERO CLUB OF AMERICA
INTERNATIONAL
AVIATION TOURNAMENT
Employee
BELMONT PARK 1910
Vancouver BC
Canada
July 31-1913
3PM.
Alys McKey
Bryant
EB

taking FLIGHT

FROM THE NINTH century onward, Middle Eastern and European documents record proposals for flying machines and even occasional attempts at flight by rash individuals using artificial wings. Leonardo da Vinci (1452–1519) sketched and built ornithopters, model aircraft that were propelled by flapping wings. Nevertheless, humans remained earthbound until November 1783, when the Montgolfier brothers demonstrated their hot-air balloons in Paris. Jean-François Pilâtre de Rozier (1756–1785) and the Marquis d'Arlandes (1742–1809) became the first humans to ascend in a balloon in free flight, making a twenty-five-minute journey and landing some 8 kilometres (5 miles) from their take-off point.

Ballooning was chiefly a form of public entertainment, but gradually found military application as a platform from which to observe artillery fire. Balloons were also used in pioneering exploration of the upper atmosphere. In Canada, they were almost exclusively used for entertainment, ascents being made by visiting foreigners, often sporting fictitious academic titles. On August 10, 1840, Frenchman Louis Anselm Lauriat (1785–1858) made the first ascent from Canadian soil at Saint John, New Brunswick. In September 1856 another French citizen, Eugène Godard (1827–1890), constructed a balloon in Montréal—Canada's first flying machine—and made several flights with a few passengers above the city.

facing page: These early flight artifacts include J.A.D. McCurdy's scrapbook (bottom left), an aviator's certificate (Canadian pioneer William Stark, bottom centre), the *Silver Dart*'s propeller (centre), and experimental propeller models made by Alexander Graham Bell.

facing page: Europeans witness the wonder of flight, in this case a Borel Morane monoplane, *c.* 1911.

As ascents became more common, promoters added airborne trapeze acts and parachute descents to maintain public interest. Most performances went smoothly; crowds paid admission fees to fairgrounds or racetracks, watched intently as the balloon was inflated, and cheered lustily as the aeronaut floated up—and the aeronaut usually alighted safely several kilometres away. Occasionally, something went wrong. On September 26, 1888, several volunteers were assisting a balloonist at the Central Canada Exhibition in Ottawa. One of them, Tom Wensley, held on to the ropes too long, lifted off with the balloon, then fell to his death—the first aeronautical fatality in Canada.

Balloons were important, even if they represented a technology with limited prospects. Just as steam locomotives demonstrated that humans could travel at great speed without having their lungs crushed, balloon ascents demonstrated that flight was not necessarily fatal. Experimentation in all fields became less terrifying. In aeronautics, research proceeded along two lines: lighter than air (refining the balloon) and heavier than air (leading to the airplane).

The former involved making balloons in a navigable shape (roughly a cigar), fitting a framework for seating and controls, and then adding a power unit. In September 1852, Henri Giffard (1825–1882) flew an airship from Paris to Trappes, some 27 kilometres (17 miles). His craft was driven by a lightweight steam engine. Although at least one airship used an electric motor, the future belonged to the newly invented gasoline engine.

HEAVIER-THAN-AIR FLIGHT: THE LONG QUEST

Success with heavier-than-air flight took longer. The most original, practical, and significant figure in the early history of manned flight was an English country gentleman, Sir George Cayley (1773–1857). From youthful flying toys and the scientific observation of birds, he identified and lucidly described the basic forces of flight: lift, drag, and thrust. His article "On Aerial Navigation," published in parts in 1809 and 1810, launched the quest on rational lines. Cayley dabbled in the theory of manned flight and appreciated the need for a light, practical engine. He was consulted by other pioneers, and late in life he built a crude glider (but had his coachman test it). However, his early writings were far more significant than his later experiments.

All over the Western world, people began to study the problems of flight. The first wind tunnel was used in England in 1871. Established scientific societies and new groups specializing in aeronautics began to study the topic. What was probably the first international conference on aeronautics was held in Paris in 1889; a meeting convened in Chicago in August 1893 was attended by an array of cranks as well as scientists. Finally, an international body, the Fédération Aéronautique Internationale, was created in Paris in October 1905.

Two figures stand out in this period of early experimentation: Otto Lilienthal (1848–1896) and Octave Chanute (1832–1910). Lilienthal built his first glider in 1890 in Germany. It was a failure, but in the next six years he developed one design after another. Some he replicated and sold. He built a craft with a motor but never flew it under power. He sought means of control other than shifting his weight around the glider but never went beyond the design stage. Nevertheless, propelling himself by running down hills, he repeatedly soared short distances—making some 2,500 flights—though he was never able to get higher than his initial launch point. He preached in writing and by example that the best way to learn to fly was by flight itself: trial, error, and practice. It was bold, but it proved fatal; Lilienthal died on August 10, 1896, from injuries received after stalling and crashing on his favourite hill the previous day. The headstone on his grave was inscribed *Opfer müssen gebracht warden!* ("Sacrifices must be made!").

Chanute never flew but was important as an information broker. Writing widely, he informed aviation pioneers of all that had been done in the past and defined what had to be done next. He transmitted information back and forth between Europe and North America. Such was his prestige that he gave the opening address at the Chicago conference, where he spoke of "the fascinating but unsolved problem of aerial navigation." Chanute promoted the Lilienthal-type hang-glider in biplane form as the most promising design. Above all, he befriended and encouraged a wide range of men dabbling in aeronautics, including the Wright brothers.

Pioneers worked with private funds. Some toiled in relative secrecy, others in the full glare of publicity. When Wilbur and Orville Wright announced their successful first powered flight in December 1903, the scientific community was skeptical. The Wrights filed patents and improved their designs. By January 1905, they had managed to fly a short circular course. However, they tended not to publicize their activities except through registration of successive patents. These were couched in general terms that led to considerable litigation as it was decided whether the designs of other pioneers were original or merely advanced variations on Wright ideas. This was especially so in the realm of aircraft control.

By 1908, pioneers in Great Britain, France, and the U.S. had achieved powered flight of a sort: brief hops, minimal manoeuvring, and landings that risked becoming crashes at any moment. Even light winds would postpone flights and complicate control. The earliest European flights were very modest compared to the Wrights' achievements. Alberto Santos-Dumont became the toast of Paris through two flights—the first on September 13, 1906, when a take-off run of 149 metres (489 feet) culminated in a 7-metre (23-foot) hop about half a metre (20 inches) off the ground, and another on November 12, 1906, when he flew a distance of 220 metres (720 feet). In March 1907, Charles Voisin made six short flights, their duration brief but the frequency notable. On September 30, 1907, Henri Farman took off on an 80-metre (262-foot) flight after 225 failed attempts, and on October 26, 1907, he flew 770 metres (2,530 feet). Also in October 1907, Robert

facing page: Wilbur Wright's sustained and controlled flight demonstrations during his visit to France in 1908 revolutionized aviation in Europe. Pictured is a flight made by Count Charles de Lambert, a Russian subject, in a Wright Flyer the following year, when he flew around the Eiffel Tower.

facing page: J.A.D. McCurdy at the controls of the *Silver Dart*. The aircraft's propeller and engine are among a small number of McCurdy artifacts in the Museum's collection.

Esnault-Pelterie demonstrated a control stick that enabled the pilot to manoeuvre with one hand, freeing the other to manipulate the throttles. However, not until January 1908 did a European aeroplane (as it was universally spelled at that time) prove capable of flying a triangular course, coming back to the starting point (which the Wrights had achieved three years earlier), and then only for 1 kilometre (less than two-thirds of a mile) total distance flown.

CANADIAN PIONEERS

The first original Canadian contributions to aviation appear to have been "improvements to flying machine" and a device "for the improvement of aerial paddle wheels" patented in 1878–79 by Richard W. Cowan and Charles Pagé of Montréal. Their devices were intended to improve control over lighter-than-air vehicles. The first scholarly discussion of flight in Canada took place at a meeting of the Association for the Advancement of Science held in Toronto on August 29, 1889, which was attended by Chanute. Six years later, on February 20, 1895, C.H. Mitchell delivered a paper titled "Aerial Mechanical Flight" to the Engineering Society in the School of Practical Science at the University of Toronto, the first such presentation by a Canadian in Canada. A pioneering Canadian, Wallace Rupert Turnbull (1870–1954), constructed a wind tunnel in 1902 at Rothesay, New Brunswick, and in March 1907 published "Researches on the Forms and Stability of Aëroplanes," the first results of his work, in the journal *Physical Review*.

Alexander Graham Bell (1847–1922) was a colossus in the national context and a major presence in North America's quest for flight. He brought to the game his intellect, enthusiasm, and encouragement; he also provided a great deal of money. Like Chanute, he was too old to fly himself. His work was thus realized through a remarkable team of young associates: Canadians Frederick Walter "Casey" Baldwin (1882–1948) and John Alexander Douglas McCurdy (1886–1961), and Americans Glenn Hammond Curtiss (1878–1930) and Thomas Etholen Selfridge (1882–1908). Bell, a Scot by birth, commuted easily between the United States and Canada. Mabel Bell, his American wife, encouraged and financed the Aerial Experiment Association.

Bell began to be interested in flight in 1892. He approached the subject by way of kite designs—box kites, round kites, and ingenious tetrahedral kites—both tethered and free.

J.A.D. McCurdy (1886–1961)

McCurdy's contributions to the development of manned flight were a prime factor in the birth of North America's aviation industry. He had also become one of the world's most experienced pilots by the time he ceased active flying, in 1916, due to a slight vision defect. However, a distinguished aviation career followed that culminated with his appointment as assistant director of Aircraft Production for the Canadian government during the Second World War. In 1948 he was appointed lieutenant governor of Nova Scotia.

On August 27, 1910, McCurdy became the first to transmit a wireless message from an aircraft, flying a Curtiss Pusher with a telegrapher's key mounted on the steering wheel and making radio contact with a ground station at Sheepshead Bay, New York. Although the range was short—2 kilometres (just over a mile)—and the message was in Morse code, its potential importance was clear.

On January 30, 1911, McCurdy set out on a daring flight from Key West, Florida, to cross the shark-infested waters of the Florida Strait to Havana, Cuba. U.S. Navy warships were available to provide assistance if needed and all went well until he was forced to land in the sea 1.6 kilometres (1 mile) short of Cuba. A crack in his engine's crankcase had caused a crippling loss of oil. Although short of his goal, he had made the longest over-water flight to that date. At a later state dinner, McCurdy was presented with a silver cup and a ribbon-bedecked envelope ostensibly containing $10,000 in prize money. Later, in his hotel room, he discovered that it contained only torn pieces of newspaper. Evidently his landing a mile off shore disqualified him from the record, but the officials had gone through the motions of awarding the prize money anyway.

Frank H. Ellis's Models of Early Aviation Experiments

WHEREAS THE contributions of Alexander Graham Bell's Aerial Experiment Association are widely recognized in the mainstream of world aviation history, most early Canadian experimenters worked in relative obscurity, often in areas of the country that lacked an industrial tradition. Knowledge of their work never became widespread, and the only reward these early pioneers received for their efforts was the satisfaction of having tried their best. These aviation pioneers lacked many things, including money, modern aeronautical data, and lightweight aero engines. During the 1940s, Frank Ellis created models to commemorate these pioneering efforts. A pioneer airman himself, Ellis is considered one of Canada's first aviation historians.

facing page, top: Underwood Brothers Oval Wing, 1908. As the news spread of the success of the Wright brothers, the sons of John Underwood, inventor of the Underwood disc plough, briefly initiated Alberta's first aircraft experiments in 1908. Lacking a suitable engine, their efforts proved unsuccessful.

facing page, bottom: Templeton-McMullen Biplane, 1909–11. William and Winston Templeton and their cousin William McMullen, of Vancouver, B.C., constructed an aeroplane that was covered with rubberized silk and mounted a British-built, three-cylinder Humber aero engine at the front. However, its power was not sufficient to keep the machine aloft.

above: Members of the Aerial Experiment Association: (left to right) Casey Baldwin, Thomas Selfridge, Glenn Curtiss, Alexander Graham Bell, J.A.D. McCurdy, and Augustus Post of the Aero Club of America, 1908.

By 1902, he had resolved to build a kite capable of lifting a human; his design succeeded on several occasions. In December 1907, using a fast boat to tow a kite, he managed to lift Selfridge aloft. Later attempts to fit his kites with engines failed.

The Aerial Experiment Association (AEA) was established in October 1907 and formalized Bell's partnership with his younger associates, each of whom brought a different talent or contribution to the quest. Baldwin (whose association with Bell was the most enduring) was a young engineer fascinated with flight. He was brought to Bell's attention initially as someone who could assist in kite designs. Selfridge, a United States Army engineer, was widely read on international experiments and was knowledgeable about meteorology. Curtiss was an expert in lightweight gasoline engines. McCurdy, another young engineer, had known Bell as a childhood neighbour and Baldwin as a university colleague. He is most often remembered as a successful and practical pilot, but equally significant was that he brought Bell and Baldwin together.

Eventually they tested Bell's towed and powered kites, but these complex designs were aerodynamic failures. The AEA had greater success with a series of remarkable biplanes:

Drome No. 1, *Red Wing*, was first flown at Lake Keuka, New York, on March 12, 1908, and piloted a distance of 97 metres (319 feet) by Baldwin, who thus became the first Canadian to fly a powered aircraft (though on foreign soil). *Red Wing* was notable for its primitive stabilizer and elevator controls.

Drome No. 2, *White Wing*, flew on May 18, 1908, over a distance of 85 metres (279 feet) by Baldwin at Hammondsport, New York. It was subsequently flown by Selfridge, Curtiss—who managed 310 metres (1,017 feet)—and McCurdy. It is notable for its hinged wingtips, which provided lateral control.

Drome No. 3, *June Bug*, was a larger version of *White Wing* and was flown on June 21, 1908, by Curtiss, who achieved distances at Hammondsport of up to 386 metres (1,266 feet) that day and even longer flights on subsequent days. It gained extensive publicity as the first aeroplane in North America to fly publicly more than a kilometre, and generated the first patent litigation from the Wrights. Adapted to primitive floats, it became the *Loon*.

Drome No. 4, the *Silver Dart*, is identified with McCurdy, who was responsible for its development within the AEA team, using lessons learned by experience with the first three aircraft. It had non-porous fabric, enlarged ailerons, and a steerable wheeled undercarriage. McCurdy gave it an initial flight at Hammondsport on December 6, 1908, followed by several more flights over varying distances, including one of over 1.6 kilometres (1 mile). It was duly packed and moved to Baddeck, Nova Scotia, where on February 23, 1909, he flew it some 800 metres (half a mile) at an altitude of about 9 metres (30 feet) and effected a successful landing—the first powered, heavier-than-air flight in the Dominion.

facing page: Robert Bradford's painting *A.E.A. Silver Dart* depicts the first powered, controlled, and sustained flight in Canada, near Dr. Bell's summer home at Baddeck, Nova Scotia, on February 23, 1909.

above: A.E.A. Drome No. 3, *June Bug*. Its pilot, Glenn Curtiss, was the first in North America to fly more than a kilometre in public, winning the Scientific American Trophy for his flight at Hammondsport, N.Y., in 1908.

Tragically, Selfridge, a founding member of the AEA, was killed in the line of army duty while flying with Orville Wright on September 17, 1908. Also at that time, Curtiss was being offered independent development and marketing resources. Bell, who had envisaged the AEA as having the limited objective of achieving powered flight, concluded

William W. Gibson (right foreground) and his *Multi-Plane.* A dozen men stand on the aircraft's structure to demonstrate its strength during the "Made in Canada" Exhibition at Vancouver in 1911.

that another form of organization was needed to exploit the commercial possibilities of flight. The surviving members therefore mutually agreed to dissolve the AEA on March 31, 1909.

It was succeeded by the Canadian Aerodrome Company (CAC), headed by Baldwin and McCurdy and generously assisted by Bell. The CAC and Bell had a curious, symbiotic relationship. The young directors were also research assistants to Bell and operated in the same buildings that had housed the AEA. Nevertheless, the Canadian Aerodrome Company had its own staff, kept separate books, and pursued commercial rather than scientific objectives. They began with the old *Silver Dart,* then constructed refined versions (known as Baddeck Nos. 1 and 2). These made frequent flights (Baddeck No. 2 made one of 25.7 kilometres/16 miles), but they often disappointed on public occasions, such as the 1909 Petawawa Trials and the 1910 Montreal Air Meet.

Prodded by curious officers and by Bell, the Department of Militia and Defence consented to hold aircraft trials at Camp Petawawa, Ontario, in late summer 1909. Baldwin and McCurdy were given an opportunity to flaunt their latest creations. Unfortunately, the Petawawa trials proved disastrous. On August 2, before sunrise, with McCurdy and Baldwin taking turns as pilot and passenger, they made three short flights. McCurdy was at the controls on a fourth hop just as the sun came up, full in his face as he was skimming along at an altitude of 3 metres (10 feet). He misjudged his clearance, struck a knoll, and crashed. Both men were scratched and bruised, but the *Silver Dart* was a write-off.

Their hopes lay in a successful demonstration of their second machine. Baddeck No. 1 was shipped to Petawawa as a replacement. Unfortunately, it was untried and neither pilot was completely familiar with its characteristics. On August 12, McCurdy made

Jacques de Lesseps (1885–1927)

THE STAR of the 1910 Montréal and Toronto shows was Count Jacques de Lesseps, son of the builder of the Suez Canal. Suave and skillful, he was a natural showman. Montréal City Council received him at a special session on June 27, even before he performed his most spectacular acts. The first newspaper advertisements for the Toronto meet featured his name and picture to the exclusion of all others. Following his flight over Toronto on July 13, de Lesseps was draped in a tricolour and carried shoulder-high before a grandstand of madly cheering spectators. De Lesseps's appearances were just the beginning of his Canadian associations. He married the daughter of a Canadian railway baron, and in 1926 he returned to Canada in the employ of La Compagnie Aérienne Franco-Canadienne. On October 18, 1927, he and his mechanic were killed while flying in bad weather near Ste. Félicité, Québec.

an unimpressive 90-metre (300-foot) hop before an official party. The next day an even shorter flight ended as the tail-heavy aircraft crashed. The damage was less severe than that to the *Silver Dart,* but neither aircraft had put on a good performance.

Several officers repeatedly recommended that funds be allocated to supporting experiments at Baddeck or purchasing two aeroplanes. The cabinet occasionally hinted at keeping an open mind on military flying, but refused to vote a penny towards it. With no sales and the Canadian government uninterested in aviation, the CAC was doomed as a commercial venture. McCurdy in particular despaired at their lack of success, which was underlined as more advanced machines appeared in Canada. The CAC was never formally dissolved but died quietly towards the end of 1910 as Bell, Baldwin, and McCurdy went their separate ways.

Before the First World War, there were isolated attempts at flight by various individuals. Laurence Jerome Lesh, an American living in Montréal, tested a glider in mid-August 1907 and persisted in his research until 1911. Robert McCowan of Sydney, Nova Scotia, spent six years in aeronautical research and built a glider in 1910, but, like Lesh, did not experiment further. Other Canadians pursued their own designs, but these pioneers were defeated by lack of funding and sometimes by communities that ridiculed their efforts.

Among those who fell short was William Wallace Gibson (1875–1965). First in Saskatchewan, then in British Columbia and Alberta, he tried to balance family obligations against his thirst to fly, all the while being treated with indifference by his neighbours. He began with kites, which became ever larger and more complex. When the Wrights' first successful flights were reported, Gibson renewed his work, starting with an engine and propeller design. That occupied him until 1910, when he married it to an awkward machine—the Gibson *Twin Plane*—that flew some 60 metres (200 feet), with scarcely any control, and crashed. Starting over again, he designed a new machine, the

Multi-Plane, which made a few hops but was demolished in a crash in 1912. With no backer comparable to Alexander Graham Bell, Gibson abandoned his experiments.

Aerial exhibitions in the era of early powered flight began at Rheims, France, in August 1909, drawing their inspiration from the great international expositions and world's fairs of the nineteenth century.

In 1910, the first Canadian air meets were held at Lakeside near Montréal, then at Weston near Toronto. Both generated excitement but lost money. The Lakeside meet was a spectacular event, even by the standards of the day. It incorporated balloon ascents, parachute descents, dirigible flights, and flypasts by an array of aircraft. Some failed to get airborne, one dirigible was destroyed, and three aircraft crashed, fortunately without causing serious injuries.

Pre-war exhibition flying in Canada followed a pattern. In 1910, the major cities of Vancouver, Montréal, Toronto, and Winnipeg held public demonstrations. By 1911, there *facing page:* The oldest surviving Canadian-built machine to attempt flight in Canada, the Museum's McDowall Monoplane was built on a part-time basis. Typical of the efforts of early Canadian amateur experimenters, it was not completed until 1915, by which time its design had long since been superseded.

below: Although it is uncertain whether the McDowall Monoplane actually achieved flight, with wings removed it doubled as an ice scooter, *c.* 1920.

were enough pilots and machines that aircraft were introduced at moderately sized cities such as Hamilton, Calgary, Edmonton, Lethbridge, Saskatoon, Victoria, Ottawa, Regina, and Québec. They continued to appear at these centres in 1912, but by then even small communities like Napanee in Ontario, Sherbrooke in Québec, Brandon in Manitoba, and Nelson in British Columbia, to name a few, were visited. Many appearances coincided with local fairs or provincial agricultural exhibitions. The early barnstormers had to contend with the hazards of improvised landing grounds—such as gopher holes, panic-stricken livestock, and enthusiastic spectators rushing the field as the aircraft touched down. Almost all these performers were foreigners; J.A.D. McCurdy was the only Canadian pilot who routinely followed the air-meet circuit.

The aviation meets had far-reaching effects. Here and there people attempted to build aircraft patterned on what they had seen. A few men, later famous in Canadian aviation, were first inspired by the frail machines testing the skies before crowds at these meets or at subsequent country fairs. Among these were John H. Parkin, scientist, writer, and a founder of the Canada Aviation Museum; Gordon R. McGregor, a future president of Trans-Canada Air Lines and Air Canada; William G. Barker, later Lieutenant-Colonel W.G. Barker, Canada's most decorated airman of the First World War, and Walter E. Gilbert, later a distinguished frontier pilot. Many of these young aviation enthusiasts attempted to build their own aircraft, often following the designs of others. Robert McDowall, a civil engineer and land surveyor from Owen Sound, Ontario, was one of them.

He first saw heavier-than-air flying machines during a visit to England and France in 1910, and his interest was aroused. He soon bought a used fan-type Anzani engine in New York and proceeded to design an aircraft similar in appearance to the popular Blériot XI, though he was never able to get it airborne.

facing page: Georges Mestach seated in his Borel Morane monoplane. An important addition to the Museum's collection, it is the oldest surviving aircraft known to have flown in Canada.

Before the outbreak of the First World War, only four Canadians had acquired Aero Club of America certificates, the contemporary equivalent of a licence. These were:

- Certificate no. 18, issued to J.A.D. McCurdy of Baddeck, Nova Scotia, on October 23, 1910
- Certificate no. 110, issued to William M. Stark of Vancouver, British Columbia, on April 10, 1912
- Certificate no. 132, issued to Robert B. Russell of Toronto, Ontario, on June 19, 1912
- Certificate no. 179, issued to Percival Hall Reid of Montréal, Québec, on October 23, 1912

In addition to these, a number of Canadians secured licences elsewhere. Toronto-born St. Croix Johnstone secured Royal Aero Club Certificate no. 41 in England on December 31, 1910. Jean-Marie Landry of Québec qualified for French Aviator's Brevet no. 1659 on May 30, 1914. A few other Canadians flew without obtaining formal licences. These included Tom Blakely and Frank Ellis, both then living in Calgary. In 1914 they salvaged a Curtiss biplane that had been abandoned by an American pilot, taught themselves to fly, and made a few short flights. Ellis later wrote *Canada's Flying Heritage,* a valuable history that owes much to his own participation in pioneering events.

The Canadian flying scene before the First World War was characterized by false starts and official indifference, the country's role being that of an interested spectator with modest though visible achievements in a North American context. The Aerial Experiment Association had registered several successes (notably in the design of controls) as well as singular failures (such as Bell's powered tetrahedral kites) while McCurdy had attained stature beyond Canada. Yet by 1914 the nation still had few pilots, no flying schools, and no aircraft manufacturing facilities, although several had been proposed.

It is often relatively easy to explain why something happens. It is perhaps more difficult to show why something does not happen—in this case, why Canada did not take a more active role in early aviation. Political inertia was clearly part of the equation, but other factors were present. One was a lack of engineering schools and infrastructure; Canada was only beginning to embark on basic scientific research, as opposed to technological tinkering. With few scientists or engineers to advise them, investors found it difficult to distinguish between the promising and the impractical. Industrialists and financiers were unwilling to invest heavily in Canadian aviation. It would take a world war to demonstrate how and why this should change.

The Borel Morane Monoplane

IN 1909, Louis Blériot gained worldwide fame for crossing the English Channel in his Blériot XI monoplane. Raymond Saulnier had worked with Blériot but soon decided to design and build an airplane of his own. In 1911, Saulnier formed the Société anonyme des aéroplanes Morane-Borel-Saulnier with Léon Morane and Gilbert Borel; together they developed the little monoplane known as the Borel Morane.

The Museum's example was imported into the United States by Georges Mestach, an early Belgian exhibition pilot. It is the oldest known surviving aircraft to have flown in Canada. Mestach was among a handful of Europeans to fly in North America during aviation's early years. His exhibition flying during 1912 included stops at Québec, Sherbrooke, and Winnipeg, where the Borel Morane proved no match for the stiff prairie wind; it was damaged after crashing into a fence. The machine's checkered career also included a crash at an air meet near Chicago that resulted in North America's first midair collision fatality.

Mestach sold the damaged aircraft but continued to fly for the new owner until it was purchased in 1914 by an American exhibition pilot, Earl S. Daugherty of Long Beach, California. Mestach died in the 1920s while trying to establish an airline in the Belgian Congo. Although Daugherty also suffered a fatal aerial accident in 1928, the aircraft remained in his family's possession until it was purchased by the Museum in 2002.

2

the FIRST WORLD WAR

ARMIES BEGAN EXPLORING uses for aeroplanes in 1908. Demonstrations before curious generals and war ministers were followed by the deployment of aircraft on exercises, the training of military pilots, and finally, the initial use of aircraft in warfare in Libya in 1911 (by Italy) and in the Balkan Wars of 1912–13 (by Bulgaria, Greece, and Serbia).

facing page: James Leech's painting *Attack on Zeppelin L.22* depicts the airship's destruction in 1917 by a Curtiss Large America flying boat piloted by Robert Leckie of the Royal Naval Air Service. Leckie was the only pilot ever associated with the destruction of two Zeppelins.

Military air organizations in various countries developed from different roots, yet in similar ways. In Germany, the Transport Corps fostered aviation; in the United States, the Signal Corps did the same. In Britain, an Air Battalion was organized within the Corps of Royal Engineers, which in 1912 became a special army branch, the Royal Flying Corps (RFC), including a naval section. The latter quickly evolved into the Royal Naval Air Service (RNAS), and by 1914 the RNAS and RFC had separated, only to be reunited in April 1918 as the Royal Air Force (RAF). When the First World War began, air fleets varied in size. France possessed about 1,500 aircraft for military training and operations, Germany had 1,000, and Britain had 179.

OBSERVERS, FIGHTERS, AND BOMBERS

Prior to the First World War, the various air arms concentrated on discovering what their machines could do, how they could co-operate with other service arms—such as artillery and cavalry—and what types of aircraft were needed for the tasks at hand. One *facing page:* The Royal Aircraft Factory BE.2C was initially a successful reconnaissance machine, but as tactics of aerial fighting evolved, its inherent stability, and the observer's limited field of fire from the front cockpit, made it difficult for crews to defend themselves.

consequence was that, at the beginning of the war, the nascent air forces had a bewildering array of machines. Early in August 1914, the RFC flew to France 41 aircraft representing five different types of machines, ranging from a single Blériot Parasol to 23 BE.2a machines.

Experience from limited wars and military manoeuvres had suggested that airplanes would be most important as scouts, reporting on enemy troop dispositions and movements. This was borne out in August 1914 as German armies invaded Belgium and France. They appeared unstoppable, but an aerial report alerted British and French commanders to an opportunity to mount a counterattack. The Germans' momentum was halted, and the front stabilized into a stalemate, with opposing armies digging a series of trenches in positions that changed little over the next four years. This fortified system, bristling with barbed wire and machine guns, proved impenetrable to traditional methods of reconnaissance—the cavalry—and so airplanes assumed greater importance as scouts. The fixed front also meant that artillery was stationary; aircraft now became the means of directing gunfire on enemy positions. The BE.2c represents this army co-operation role. The original BE.2 that entered service in 1913 was designed for reconnaissance, and the BE.2c continued in action until 1917, its great stability being an asset when photographing trench systems.

But how were the airmen and soldiers to communicate with each other? In 1914 reconnaissance pilots either flew back to army headquarters with verbal reports or dropped scribbled messages to the troops. These tactics were succeeded by flare pistols, cloth panels cut into coded strips and laid out on the ground, signal lamps, and wireless radio communications using Morse code. The pioneering radios were bulky, and for most of the war airplanes could send but not receive messages. Even in 1918, two-way radio communications were rare and unreliable.

Reconnaissance pilots and observers performed varied tasks and faced diverse hazards. Flying over enemy trenches at 1500 metres (5,000 feet) was common when photographing or merely noting their defences, but mapping trench systems took crews as high as 5000 metres (16,400 feet), where cold and lack of oxygen made life miserable. More dangerous were the "contact patrols," when aircraft descended to as low as 50 metres (165 feet) amid rifle and machine gun fire to locate friendly troops and report the presence of enemy soldiers. Having no parachutes, the men feared being set on fire in the

air. In 1918, German fliers were issued parachutes for the first time; they usually worked, but sometimes did not. At the war's end, the Allies were still researching parachutes that could be worn in the tight confines of an aircraft. At this time, it was official British army policy that parachutes were not to be issued to pilots, although bulky parachutes were made available to crewmen of observation balloons, since space was not a problem.

As observation aircraft became the "eyes of the army," opposing forces sought to shoot down enemy airplanes while protecting their own. From this situation the modern fighter aircraft evolved. In 1914 pilots occasionally shot at one another with revolvers and carbines. By 1915 they carried machine guns, usually fired by an observer while the pilot concentrated on flying. Pusher-type aircraft like the de Havilland DH.2,

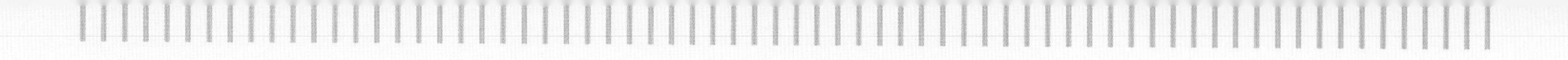

Raymond Collishaw (1893–1975)

RAYMOND COLLISHAW'S air force career was a tribute to adventure, courage, and daring. The top naval air ace of the First World War, the flyer from Nanaimo, British Columbia, remained in the Royal Air Force until the 1940s, eventually attaining the rank of Air Vice-Marshal. Collishaw was recognized as a magnificent leader, whether in command of a flight, a squadron, or an air force.

However, it was during 1917 that Collishaw rose to prominence while commanding the famous all-Canadian "Black Flight" of 10 Squadron, Royal Naval Air Service. Equipped with the highly manoeuvrable Sopwith Triplane, it was one of the most successful and feared fighting units of the time. The arrival on the Western Front of its distinctive aircraft, with black-painted engine cowlings and wheel covers and with names like *Black Maria, Black Death,* and *Black Prince,* was a nasty surprise to their opponents, the skillful and accomplished pilots of Baron Manfred von Richthofen's "Flying Circus." Always on the offensive, the squadron downed seventy-nine German aircraft in a three-month period.

Collishaw had many alarming escapades in the air and one in particular underscores the danger of open cockpit combat. While taking violent action to avoid a mid-air collision, his seat belt snapped and he flew out into space. Frantically grabbing the centre-section struts in front of the cockpit, Collishaw hung on for dear life as he was flung about the wildly gyrating aircraft. With its engine at full power, the Triplane went through a series of breath-taking dives, climbs and spins. At one point his lower body was thrown back unto the rim of the cockpit and, with strength born of desperation, he hooked his boot around the control column and pulled it forward until the aircraft leveled out. Regaining his seat after falling for nearly two miles, Collishaw set a course for home, a much-shaken but thoughtful pilot after such an awe-inspiring experience.

with engine and propeller mounted behind the pilot, allowed for the use of a forward-firing gun.

In 1915, the Germans introduced an interrupter, or synchronization gear, which enabled pilots to fire a machine gun directly forward through the propeller arc; the device timed the gun to fire only when the propeller was not threatened. Thus equipped, the German Fokker E.I, E.II, and E.III "Eindeckers" (monoplanes) revolutionized aerial combat and gave the Germans a temporary advantage over Allied aircraft.

New technology was accompanied by tactical innovations, with Germany again leading the way. Captain Oswald Boelke developed tactics that emphasized positioning (notably, approaching a target from out of the sun), close shooting, and teamwork, first between two pilots (leader and wingman) and thereafter with larger groups. Once more, Allied airmen copied their enemies. By late 1917, formations as large as sixty aircraft on each side clashed over the front, usually breaking into smaller combats after first contact. The wise pilot attempted to remain with his leader or wingman. Baron Manfred von Richthofen (1892–1918)—the "Red Baron"—paid with his life when he engaged in a solitary chase of an enemy at low level and was shot down; whether he was shot down by Arthur Roy Brown, a Canadian pilot, or by Australian machine gunners on the ground, is still disputed.

Wilfrid Reid "Wop" May as a young RFC pilot in 1918. Manfred von Richthofen was killed while chasing May's Sopwith Camel. Captain Arthur Roy Brown, May's boyhood friend, was credited with shooting down the famous "Red Baron," and May survived to become a pioneering bush pilot in Canada's north.

Aircraft evolved so quickly that advantages frequently passed from one side to the other. The "Fokker scourge" of 1915–16 was countered by Sopwith Pups; then the Germans regained a temporary edge with the twin-gunned Albatross D.IIs and D.IIIs, to be countered in turn by the Sopwith Camel and the Royal Aircraft Factory SE.5a. Types deficient in one respect might be superior in another. The Fokker DR.I triplane was relatively slow compared to Allied aircraft but exceedingly manoeuvrable when turning and climbing. In the summer of 1918, the Fokker D.VII was considered a particularly deadly opponent, yet Sopwith Dolphins and Snipes, as well as Spad XIIIs, were deemed at least as good. Superior pilots could make the best of almost any machine.

facing page: These artifacts from the First World War include William A. "Billy" Bishop's miniature medals and a model of a Nieuport 17 made for him by his ground crew (lower left).

The earliest aerial bombs were hand grenades and steel darts hurled randomly at armies below. With improved, more powerful engines, aircraft could carry greater loads. By 1916 fighter aircraft often carried a few 9-kilogram (20-pound) bombs for ground attack, and by 1918, strategic bombers routinely dropped 50-kilogram and 250-kilogram (112- and 550-pound) bombs. Handley Page o/400 aircraft carried 750-kilogram (1,650-pound) bombs, larger even than the bombs carried by the RAF in 1939. The heaviest bombs ever dropped in the First World War were 1 000-kilogram (2,200-pound) weapons carried by four-engine Zeppelin-Staaken Riesenflugzeug or "Giant" aircraft, which were as large as most Second World War heavy bombers. A smaller design was the AEG G.IV bomber, which was less formidable (and hence less famous) than the Gotha and Friedrichshafen aircraft that often raided Britain in 1917–18. With a bomb load of 400 kilograms (880 pounds), it was limited to tactical bombing near battle fronts, but it was also used for long-range reconnaissance and photography. Its crew of three or four men could change positions during flight should the need arise.

Although most bombing raids occurred over or near the battlefield, in 1915 German Zeppelin airships began bombing cities, and the Allies retaliated with their own attacks. Both sides claimed to be attacking industrial or military targets, but aiming was primitive and so most bombs missed any significant target. Although civilian casualties were low, the psychological effect of bombing was impressive; a daylight raid on London by Gotha bombers on June 13, 1917, panicked the population. More than twenty years later, "strategic bombing" was to assume its full, terrible potential.

Canadians were present in the wartime bombing campaigns, both as defenders and as attackers. At least six German Zeppelins were shot down by Canadians or men with strong Canadian connections. Captain William W. Rogers of Alberton, Prince Edward Island, was the first person to shoot down a Gotha over France, on December 12, 1917.

In 1916–17, No. 3 (Naval) Wing, attacking targets in western Germany, included at least thirty-seven Canadians flying Sopwith 1½ Strutters and Handley Page o/100 bombers. Subsequent squadrons usually had a sprinkling of Canadians. One of these, Colonel Redford Henry Mulock of Winnipeg, commanded a force of Handley Page v/1500 bombers that would have bombed Berlin in November 1918 had not the Armistice been signed.

ROYAL AIR FORCE
TECHNICAL NOTES
HEIGHT
W.A.B.
ROYAL FLYING CORPS
NOTES ON RIGGING FOR AIR MECHANICS
F

R
8239D

Airplanes assumed other roles. In the Mediterranean theatre, RNAS pilots (some of them Canadians) dropped torpedoes aimed at Turkish warships. Over the sea, both sides used airships and airplanes to seek out hostile fleets. (An RNAS aircraft made the first contact with the German fleet at the Battle of Jutland.) German submarine warfare led the British to employ aircraft to escort convoys and to hunt and bomb U-boats. Several submarines were reported sunk, but aircraft seem to have destroyed only one such craft: an Austrian submarine was bombed by an Italian flying boat in the Adriatic. Nevertheless, convoys under aerial protection were less prone to submarine attacks.

Notable Canadian anti-submarine pilots included Lieutenant Stuart Graham and Flight Lieutenant Basil Hobbs, both of whom would pioneer flying boat operations in Canada between the wars. In France and Mesopotamia, airplanes dropped supplies to isolated army formations. Indeed, in the First World War airplanes performed almost every task that they would in the Second, with the exceptions of aerial evacuation of casualties and transportation of troops (although individual spies were sometimes parachuted behind enemy lines).

Canadians flew in all theatres of war—the North Sea, Britain, France, Italy, the eastern Mediterranean, East Africa, Palestine (assisting Lawrence of Arabia, among other tasks), and Mesopotamia. They manned observation balloons and piloted airships. When hostilities ended elsewhere in November 1918, Canadian airmen were still engaged with British units supporting anti-Bolshevik forces in what had become the Russian civil war.

facing page, top: Canadians flew in all theatres of the First World War. Captain C.M. "Black Mike" McEwen piloted a Sopwith Camel in Italy and was credited with 27 victories. As an air vice-marshal in the RAF during the Second World War, he commanded Bomber Command's 6 Group (Canadian).

facing page, bottom: Less formidable than the larger Gotha bombers that raided the civilian populace of Britain, the AEG G.IV was limited to tactical bombing near the battle front and, depending on the mission, the crew consisted of three or four men.

above: The Museum's AEG G.IV is the only twin-engine German bomber of the First World War still in existence.

Canadians had not been present at the very beginning of aerial warfare but were there in strength as the first major air war concluded.

The war affected military theory and thinking. Some officers became convinced that mobile armour was the ultimate future weapon; airmen like Hugh Trenchard in Britain and William Mitchell in the United States were equally certain that air power would trump every other consideration in future wars. Many believed that aircraft would play much the same part in another war as they had in the one just past—as an adjunct of the army. In fact, no one could foresee the enormous changes still ahead—from streamlining to radar—that would reinforce some theories and negate others.

Wartime demands pushed social as well as technical development. The Royal Air Force ultimately organized a uniformed women's branch; its members performed traditional "women's" roles but also drove vehicles and serviced aircraft. This liberal approach did not extend to Canada, where the RAF operated a training program in 1917–18. Although some 1,200 women had clerical and mechanical jobs there, they were employed as civilians only. In May 1918, the RAF considered recruiting for a Canadian-based Women's Royal Air Force, but dropped the idea because the cost of barrack accommodation for women was estimated at $430 per capita versus $235 per capita for men. Probably the largest Canadian female presence in the air war was the workforce employed in manufacturing by Canadian Aeroplanes Limited of Toronto.

above: The use of metal for aircraft skins was pioneered by Germany's Hugo Junkers during the First World War. Corrugated aluminum is evident on the Museum's Junkers J.I., an armoured ground attack aircraft that arrived in Canada as a war prize in 1919.

facing page: Portrait of Robert Leckie, taken during the Second World War. He is wearing the insignia of an air vice-marshal, but was promoted to Air Marshal and Chief of the Air Staff in 1944.

The most obvious changes had been in the realm of technology. Range, speed, and load capacity had all risen. Engine horsepower increased fivefold during the war and pushed service ceilings above 7 200 metres (23,600 feet). More significant, however, was that engine reliability had improved markedly. Henceforth, people put greater trust in airplanes to do a job and do it well. The proof was to be seen in the fact that the statesmen who commuted between London, Paris, and Brussels in the course of negotiating the 1919 peace treaties did so by air.

Robert Leckie (1890–1975)

ROBERT LECKIE'S military career began during the First World War, and on a number of occasions during the next thirty years he shaped Canada's aviation heritage. A Scottish immigrant at the age of sixteen, he learned to fly in 1915 at the Curtiss Aviation School in Toronto. Joining the Royal Naval Air Service, he operated from Great Yarmouth in the U.K. on anti-submarine patrols and gained a reputation for being able to handle the worst North Sea weather. He was the only person ever involved in the destruction of two Zeppelins. Shooting down Zeppelin L.22 in 1917 was a particularly dangerous operation in the slow Curtiss Large America flying boat he was piloting. In 1918, his de Havilland DH.4 was credited with destroying Zeppelin L.70, with the commander of Germany's Zeppelin fleet, Peter Strasser, on board, during a raid on the U.K.

After the war, Leckie served on the Canadian Air Board as director of flying operations and played a major role in the organization of the Canadian Air Force and the dispersal of Imperial Gift aircraft to Canada. In 1920, he organized and led the first trans-Canada flight, which employed flying boats from Halifax to Winnipeg and wheeled aircraft onward to Vancouver. Carrying mail from coast to coast by air for the first time, the flight served as a harbinger of the future.

Leckie served in various capacities after returning to the Royal Air Force, including command of flying operations on HMS *Hermes,* one of Britain's early aircraft carriers. In the mid-1930s, while in charge of elementary flying schools for the Royal Air Force (RAF) Reserves in Great Britain, he studied existing plans to establish a comprehensive empire flying training plan and suggested that it be established in Canada. He also came to value the newly developed Link trainer, and ordered them installed at all the RAF's elementary flying schools, thus initiating use of one of the world's first flight simulators. At this time his responsibilities included training many of the airmen who would later fight in the Battle of Britain.

Appointed in 1940 as the director of training of the Royal Canadian Air Force (RCAF), Leckie was a key figure in the implementation of the British Commonwealth Air Training Plan. His organizational skills and drive contributed to the eventual success of this mammoth air training scheme, which was possibly Canada's greatest contribution to the Allied war effort. Promoted to Air Marshal and Chief of the Air Staff in 1944, Robert Leckie retired from the RCAF in 1947.

facing page: A Burgess-Dunne seaplane was Canada's first military aircraft. Robert Bradford's painting *Burgess-Dunne* shows this aircraft, which was purchased in the United States and flown to Canada in 1914. Later shipped overseas, it was abandoned without seeing action, ending Canada's attempt to establish a Canadian Aviation Corps.

Few monoplanes had seen wartime service; biplanes, with their struts and bracing wires, had dominated the skies. The evolution of cantilever construction increased wing and fuselage strength, as did the use of the tubular metal frames that replaced wooden structures. Metal skins, pioneered by Germany's Hugo Junkers, enabled the airframe covering to bear part of the load. These developments reached their wartime peak in the Junkers J.1, an all-metal monoplane in which the absence of high-drag bracing compensated for its extra weight. These steps made possible the interwar evolution of both high- and low-wing monoplanes; the latter would, in turn, facilitate the evolution of retractable undercarriages.

TOWARDS A CANADIAN AIR FORCE

It is not surprising that the major powers had taken an early and active interest in military aviation; in most cases, they had previous experience with observation balloons in either warfare or peacetime manoeuvres. Nor is it surprising that a few small nations such as Serbia formed aviation corps as early as 1911; they were engaged in "hot" wars and were ready to grasp any new weapon offered. Prior to the war, the Canadian Militia had mostly ignored the subject. Even a 1912 visit to Britain—where the Royal Flying Corps was already in existence—did not sway the then-minister of militia, Sam Hughes. He was quoted as saying: "You will get just as good results by climbing a mountain or a church steeple." A year later, a reporter stated that Hughes regarded aviation as "of little use in modern warfare" and "a useless expense to Canada." With the minister of militia setting the tone, the federal cabinet's view of flying was that of massive indifference verging on hostility.

In August 1914, the month in which war began, Sam Hughes forgot his previous indifference and authorized the formation of a Canadian Aviation Corps (CAC). He commissioned Ernest L. Janney of Galt, Ontario, to lead the force. Janney did not even know how to fly, but he bought a Burgess-Dunne floatplane in the United States and arranged for a company pilot to fly it to Québec in time for sailing with the 1st Division of the Canadian Expeditionary Force (CEF), the first Canadians to proceed overseas. The seaplane, built by the W.S. Burgess Company of Marblehead, Massachusetts, and based on a design by British aerial pioneer Lieutenant J.W. Dunne, was stable but unmanoeuvrable, so was difficult to fit into any military role. While in Québec Janney recruited another officer,

Lieutenant William Sharpe, who had taken flying lessons in the United States but did not have a pilot's certificate. Once in Britain, the airplane—Canada's first military aircraft—disintegrated before it had even flown, Janney resigned his commission, and Sharpe went to a British flying school, where he died in a crash on February 4, 1915—Canada's first wartime aerial fatality. The CAC vanished at his death. Canada would organize no distinct air force of its own until the fall of 1918.

The first Canadians engaged in aerial warfare went overseas as CEF soldiers and transferred to the RFC later. They were frequently attached to army co-operation squadrons as observers for a few months and then trained as pilots. Throughout the war the CEF remained a source of airmen transferring to the RFC, but other Canadians entered

the air services by different routes. In 1915, both the RFC and RNAS began recruiting in Canada. At first, men were compelled to obtain Aero Club of America certificates at their own expense at private schools (most notably the Curtiss Aviation School in Toronto and the Wright School at Dayton, Ohio). Having earned their certificates, they were reimbursed by British authorities, granted provisional commissions, and posted to Britain for advanced training. The commercial schools could not cope with the number of candidates applying, and eventually both air services waived the requirement for an Aero Club of America certificate. Continued RFC expansion, coupled with heavy casualties, led to the inauguration in 1917 of the "RFC Canada" scheme (see chapter 10).

We do not know precisely how many Canadians joined the First World War flying services, but estimates range from 13,000 to 23,000. As of 1918, between one-quarter and one-third of all flying personnel in the RAF were Canadians. If enlistment statistics are uncertain, casualties figures are not; 1,388 Canadians were killed or died in aerial service.

above: Alan McLeod, aged 19, was awarded the Victoria Cross for piloting his burning aircraft to a crash-landing while standing on its wing to avoid the flames, and then saving his gunner from the wreckage.

facing page: In *Barker's VC Combat*, Robert Bradford depicts William Barker's single-handed engagement with one of several large formations of German fighters. Severely wounded, he crashed his Sopwith Snipe behind Allied lines and was later awarded the Victoria Cross for his actions.

There had also been formal honours accorded the Canadians. Approximately 800 were awarded decorations or were mentioned in dispatches. The gallantry awards included numerous foreign medals, as well as three Victoria Crosses. The first of these was awarded to Captain William A. "Billy" Bishop of Owen Sound, Ontario, who was credited with a solo attack on a German airfield on June 2, 1917, while flying a Nieuport 17. Second Lieutenant Alan A. McLeod of Stonewall, Manitoba, piloted his burning Armstrong Whitworth FK.8 to the ground while his gunner fought off German fighters, then rescued him from the wreckage on March 27, 1918. Major William A. Barker of Dauphin, Manitoba, climaxed a distinguished fighter career on October 27, 1918, when he single-handedly engaged numerically superior enemy formations and destroyed four of his opponents.

E8102

below: William A. "Billy" Bishop, seated in a Nieuport 17, adjusts the Lewis gun, which has been pulled down on its Foster mount for accessibility, *c.* 1917.

facing page: Removal of a piston and connecting rod from a Gnome Monosoupape rotary engine. As in all rotaries, the cylinders and crankcase revolved around the engine's stationary crankshaft. From the *Gnome Mono Instruction Book*, *c.* 1917.

It is impossible to accurately determine the exact number of enemy aircraft that were destroyed or damaged by individual airmen during the First World War. This is due to the varying rules applied by the various air services to define an aerial victory and the ways in which information from combat reports and other sources was interpreted or recorded. Inaccuracies aside, victory scores often attributed to pilots such as "Billy" Bishop (seventy-two), Raymond Collishaw (sixty), Donald McLaren (fifty-four), and William Barker (fifty) still attest to the skill, determination, and courage displayed by these Canadian pilots. Names occasionally emerged because of an airman's association with a particular air engagement. Wilfrid "Wop" May was Manfred von Richthofen's last intended victim during a low-level chase over the Somme River valley. Roy Brown, who knew May before the war, intervened and was credited with shooting down the famous "Red Baron," although recent research suggests that Australian ground gunners might have been responsible.

The popular concept of Canadian airmen is that they were country lads, dead shots almost by nature. Statistics show, however, they were more likely to have come from a city or large town than from the countryside and from professional or student backgrounds. The average age on enlistment (or transfer from another service) was twenty-three.

In the closing months of the war, Canada at last formed its own military air organizations at home and abroad. In June 1918, the presence of German submarines in North American waters finally moved authorities to establish two seaplane bases in Nova Scotia at Halifax and Sydney, and, in September 1918, the Royal

Canadian Naval Air Service (RCNAS) was created to operate aircraft and airships on anti-submarine patrols. However, the RCNAS had no trained personnel, so the two air stations hosted U.S. Navy aircraft and crews until Canadians could be trained. The RCNAS was still in the process of formation when the Armistice was signed on November 11, 1918, and it was disbanded in December 1918.

The Canadian Air Force (CAF), formed in Europe, had a somewhat longer life. The concept of distinct Canadian squadrons within the RAF was approved in May 1918, and in August a Canadian Air Force Detachment was formed at Halton, England, to train mechanics for two proposed Canadian squadrons. Reluctantly, the RAF set about transferring Canadians from its own units to the new organization. On November 20—nine days after the Armistice—Nos. 81 and 123 Squadrons (RAF) were redesignated Nos. 1 and 2 Squadrons (CAF), based at Upper Heyford and later at Shoreham. They were equipped, respectively, with Sopwith Dolphin fighters and de Havilland DH.9 bombers. Unlike the RCNAS in Canada, the CAF units were manned by experienced aircrews. With no war to fight, however, the officers tested their aircraft, planned for post-war developments, and ultimately supervised the packing of aeronautical equipment for shipment to Canada. No. 1 Squadron was disbanded on January 28, 1920, and the last of the overseas CAF was dissolved on February 5, 1920. A handful of Canadians remained in Britain as members of the Royal Air Force, participating in Britain's colonial wars and aeronautical progress. Most of those repatriated to Canada returned to their civilian pursuits. Enough, however, had been so entranced by flight that they sought new adventures at home.

The Rotary Engine

MOST AIRCRAFT piston engines are stationary, with a crankshaft turning the propeller. Rotary engines, by contrast, had a stationary crankshaft around which the whole engine and propeller rotated. They offered reasonable power combined with light weight and were self-cooled. The first rotaries appeared in 1908, and French designs (Gnome, Le Rhône, and Clerget) were long dominant. The gyroscopic effect of a spinning engine in fighter aircraft bestowed tremendous manoeuvrability in turns. However, when fighter tactics evolved from "dogfighting" to "hit-and-run" methods, the rotary-powered aircraft was at a disadvantage against heavier, faster diving machines. By 1918, the rotary engine had reached its peak of development.

"CITY OF WINNIPEG"
AIRLINES, LTD.

BARNSTORMERS *and* BUSH PILOTS

3

facing page: Robert Bradford's *"Doc" Oaks and Friend* portrays Harold A. "Doc" Oaks, founding manager of Western Canada Airways, standing in the open cockpit of Fokker Universal G-CAFU *City of Winnipeg,* mid-1920s.

War had forced aeronautical development into specific channels. Peace enabled it to return to other, more pacific endeavours. From Australian deserts to Venezuelan highlands, from the Arctic to the Antarctic, nations adapted and developed aviation according to their specific needs, resources, and aspirations. Progress in Canada in some ways resembled that elsewhere; often it followed patterns unique to the nation.

By the time of the Armistice of November 1918, Canada offered great potential for aviation but very few resources. The only developed air stations were the two U.S. Navy seaplane bases in Nova Scotia and six Royal Air Force training stations in Ontario. Initially, commercial operators employed former military aircraft. Despite these handicaps, many aerial veterans returning to Canada were determined to continue flying, one way or another.

At first, Canadian aviation existed in a legal vacuum. A local branch of the Aerial League of the British Empire enthusiastically publicized aeronautics, appointed itself as a reception committee for returning air force officers, and issued "Aviator's Certificates" to them without any checks on their qualifications. This ended in June 1919 when the government established the Air Board to license and supervise pilots, mechanics, aircraft, and air bases. The Air Board also planned how Canada might best exploit aviation. The Aerial League welcomed the new legislation but wryly observed: "We are sorry that we

above: Stuart Graham, Canada's first bush pilot. Entries in his logbook, now in the Museum's collection, detail the launch of Canadian bush flying by the Laurentide Pulp and Paper Company in Québec's St. Maurice Valley in 1919.

facing page: Airmen of Northern Aerial Minerals Exploration Ltd., which, along with several other pioneering companies, was the first to penetrate the high northern latitudes of Canada. In 1928, T.M. "Pat" Reid (third from left, back row) made the first aerial circumnavigation of Hudson Bay.

should have efficient regulations before we have anything to regulate." The Air Board was disbanded in 1922, its regulatory duties being taken over by the Canadian Air Force (made "Royal" in 1923). In 1936, the new Department of Transport became the regulating body for civil aviation.

After the creation of the Air Board, Canadian aviation was dominated by "barnstormers," successors of the itinerant, mostly foreign, pre-war pilots who moved from town to town, demonstrating their exotic machines. This new generation of pilots was overwhelmingly composed of Canadian veterans of the British flying services. A few had distinguished war records, but fame was no guarantee of commercial success, as demonstrated by Bishop-Barker Aeroplanes, a company founded by two Victoria Cross recipients, which nevertheless failed after a few months.

The most common barnstorming aircraft in 1919–20 were war-surplus Curtiss JN-4s. Their pilots made numerous historic flights: the first airmail flight between Prince Edward Island and the mainland, the first flight from Victoria to Vancouver (across the Strait of Georgia), and the first flight through the Canadian Rockies. Barnstormers performed limited aerobatics at fairs, raced against automobiles, dropped advertising leaflets, employed wing walkers and parachutists—anything to draw a paying crowd. Most often they carried passengers on short flights, charging a dollar per minute in the air, hoping a few customers would go on to buy flying lessons.

Air Board inspectors criss-crossed Canada, ensuring compliance with regulations and investigating accidents. Some of the more foolhardy practices ceased. The inspectors laid down maintenance standards and refused to license some deteriorating machines. Unhappily, barnstormers had difficulty finding replacements they could afford. Rising costs drove many out of the business; a few eked out a hand-to-mouth existence until they could organize a successful operation or be hired by someone who had.

AVIATION'S SEARCH FOR A CANADIAN ROLE

Barnstorming gave little clue as to how airplanes might find commercially viable roles, but astute thinkers found practical applications. Ellwood Wilson, chief forester for the Laurentide Pulp and Paper Company, decided to use aircraft to survey the company's

timber concessions in Québec's St. Maurice Valley. The U.S. Navy had left twelve Curtiss HS-2L flying boats at Sydney and Halifax, and Wilson acquired no. 1876, later registered as G-CAAC and ultimately known as *La Vigilance*. He hired pilot Captain Stuart Graham of the Royal Air Force and launched Canadian bush flying.

Graham's logbook, held by the Canada Aviation Museum, details this pioneering episode. He left Halifax with "1876" on June 5, 1919, proceeding in stages (with interruptions due to weather) to Grand-Mère, Québec, arriving on June 8. A week later he returned to Halifax to take delivery of a second HS-2L, no. 1878 (later G-CAAD). That summer these two machines and their crews made survey flights, reported forest fires, and photographed selected Laurentide properties.

All twelve U.S. Navy HS-2Ls were registered in Canada (ten with the Air Board, for use in government operations). Sturdy and adaptable, they were the nation's first significant civil aircraft. A further thirty-one were imported and used by the Air Board, Royal Canadian Air Force (RCAF), Laurentide Air Service, Ontario Provincial Air Service, Manitoba Government Air Service, and several smaller operators. Prudent crews carried an array of tools, spare parts, and messenger pigeons in case they were forced down on remote lakes. Some HS-2Ls crashed and others wore out, but the last ones continued in service until 1932. Along the way they accomplished an impressive number of Canadian firsts: first forestry patrols and surveys (1919), first mining claim staked by airplane (1920), and first scheduled air service and airmails (1924.)

Most early bush flying was closely associated with forestry surveys and fire patrols. Between 1911 and 1922, some of the most deadly forest fires in Canadian history swept through northern Ontario. When aircraft became available, provincial governments either turned to private aerial companies (as happened in Québec, commencing with

above: Curtiss HS-2L flying boats accomplished an impressive number of firsts in Canada, including the first scheduled air services.

facing page: The Museum's re-creation of Canada's first bush plane, G-CAAC *La Vigilance,* employed parts from the original Curtiss HS-2L salvaged from a northern Ontario lake, along with components from two other aircraft. It is the only HS-2L flying boat now in existence.

Laurentide Air Service) or organized their own aerial organizations, such as the Ontario Provincial Air Service (OPAS), formed in 1924. On the Prairies, natural resources were under the control of the federal government, so the Air Board and then the RCAF did the mapping, patrolling, and transporting of fire fighters. After natural resources were transferred to provincial jurisdiction in 1931, Manitoba organized an air service modelled after the OPAS; in Saskatchewan and Alberta, private companies continued the work.

In 1920 and 1921, as HS-2LS were establishing bush flying, a larger group of machines began arriving. These were the "Imperial Gift," drawn from Britain's stocks of surplus warplanes and given freely to the Dominions. Canada received 117 aircraft plus several airships (never assembled), seaplane beaching gear, portable hangars, engines, spare parts, photographic supplies, and motor vehicles. The Imperial Gift allowed the Air Board and a nascent Canadian Air Force to begin operations.

The largest portion of the gift received by Canada comprised sixty-two Avro 504K training aircraft, some of which were adapted to forestry-survey and fire-patrol work. An array of de Havilland DH.4 and de Havilland DH.9A bombers flew on diverse tasks including forestry patrols and photography. A trans-Canada flight conducted in October 1920 used an HS-2L from Halifax to Ottawa, a large Felixstowe F.3 flying boat from Ottawa to Winnipeg, and three DH.9A machines from Winnipeg to Vancouver. Nevertheless, the British machines did not hold up well in Canada's climate. Most of them had been discarded by 1928, four years before the last HS-2L was scrapped.

Much early bush flying was close to settled areas or to the railway lines linking Canadian regions. In the winter of 1924–25, commercial firms began services to remote mining communities at Red Lake, Ontario, and Rouyn, Québec. These were costly operations and few entrepreneurs had the capital to invest in newer aircraft. However, in 1926 a Winnipeg grain merchant, James Richardson, was persuaded to enter the flying business; he hired Harold A. "Doc" Oaks as company manager to run Western Canada Airways (WCA).

La Vigilance

MUSEUMS DEAL in conservation (stabilizing artifacts that might otherwise deteriorate), restoration (repair of original artifacts), and reconstruction (manufacturing parts or even entire artifacts). The Canada Aviation Museum's display featuring the country's first bush aircraft employs all three approaches.

The original G-CAAC, named *La Vigilance,* crashed in a lake near Kapuskasing, Ontario, on September 2, 1922, happily without injuries to the crew. The submerged wreckage, with its name and registration letters still visible, was found in 1968 and salvaged in the summer of 1969. The remains of the original hull were preserved as a historic artifact, but a few parts were used in a reconstructed HS-2L built between 1975 and 1986. Aside from parts of the original G-CAAC, the new *La Vigilance* incorporated main components from two other HS-2Ls and parts from several further sources. The hull was built anew, using original factory drawings and employing original materials and methods.

Oaks had vision, experience, and business savvy. He knew what airplanes could and could not do, particularly where prospecting and mining were involved. From years of experience he also knew most of the pilots and regulators in Canada's aviation industry. He made WCA a magnet for experienced personnel.

Flying a Fokker Universal and starting with freighting contracts into northern Ontario, WCA commenced operations on December 27, 1926. Its first big task involved transportation in support of the Hudson Bay Railway, then being built through northern Manitoba. In March 1927, using two Universals, WCA flew 8 tonnes of equipment and fourteen men to Churchill, at that time beyond the railway's northern terminus, making twenty-seven round trips. The company proved what frontier aerial firms could do, provided they were financially stable, wisely managed, equipped with reasonably modern aircraft, and paid their personnel regularly.

facing page: It was the northern prospecting and mining boom, uninterrupted by the Great Depression, that nursed Canadian commercial flying from the mid-1920s to the Second World War. This scene shows the Bellanca Aircruiser of Eldorado Gold Mines Ltd. and a fuel cache at Great Bear Lake, N.W.T.

above: By 1930, Canadian Airways Limited controlled most of the air transport business in Canada. Radio operator Henry Roth poses in 1933 in the doorway of the radio cabin at Cameron Bay, Great Bear Lake, N.W.T. This building was representative of the facilities to be found at many of the company's regular points of call.

A 1930 merger of WCA with some other firms created Canadian Airways, with operations in every Canadian region. Apart from freighting, the new company secured airmail contracts that seemed to offer year-round financial stability. However, the federal government—seeking to cut costs during the Great Depression—cancelled Prairie and eastern airmail service contracts in 1931–32. It was the northern prospecting and mining boom—uninterrupted by the Depression—that nursed Canadian commercial flying from the mid-1920s through to the Second World War. Canadian Airways remained the pre-eminent operator, with competition from smaller firms such as Wings Limited, Starratt Airways, Arrow Airways, Mackenzie Air Service, and United Air Transport. Yet the situation was neither lucrative nor particularly stable. Most companies lost money, year after year. Amos Airways, for example, operated from 1936 to 1940 and turned a profit only once ($301 in 1939).

The RCAF had its own version of bush flying, chiefly carrying out forestry and photographic work. There were many suggestions that the air force should abandon such tasks and leave the field to civilian firms. Group Captain J.S. Scott, who commanded the air force from 1924 to 1928, would have preferred to vacate the field of "aid to the civil power" altogether, freeing the RCAF for military operations. James A. Wilson, the most influential civil servant in Canadian aviation, realized that the idea of a military-only RCAF could not be "sold" to penny-pinching governments, but that the budget for an air force seen to be doing something useful in Canada in peace as well as war (at least part of the time) could be defended before the Treasury Board. "Aid to the civil power" served the same purpose in the 1920s that peacekeeping has served in the late twentieth and early twenty-first centuries: making defence expenditure palatable to politicians and the public. Consequently, RCAF personnel often performed tasks that had little or nothing to do with potential combat. Even so-called service flying was composed largely of flight training.

Not all RCAF civil flying could be considered "bush work." Its pilots were the first Canadians to demonstrate aerial dusting to suppress crop and forest parasites. Fisheries protection and anti-smuggling air patrols, common in the 1920s, were flown from coastal bases as urban as Halifax and as remote as the Queen Charlotte Islands. Photo work that included a steady advance into the Northwest Territories also included surveys for the future Trans-Canada Highway and irrigation systems on the drought-

above: In Robert Bradford's *De Havilland DH9A*, an Imperial Gift aircraft of the Canadian Air Force, piloted by Captain G.A. Thompson, struggles through a canyon in the Selkirk Range on its way to Vancouver during the last leg of the Trans-Canada Flight in 1920.

facing page: In *The Big Bellanca,* also painted by Robert Bradford, pioneer bush pilot Stan MacMillan oversees the loading of radioactive pitchblende concentrate into the Bellanca Aircruiser of Eldorado Gold Mines Limited at Great Bear Lake, N.W.T., during the northern prospecting and mining boom of the 1930s.

ravaged Prairies. Nevertheless, other operations were decidedly "frontier" tasks, such as the Hudson Strait Expedition of 1927–28. With the Hudson Bay Railway approaching the port of Churchill, questions persisted about when Hudson Strait would be ice-free. The fact that Western Canada Airways was airfreighting cargo to Churchill at about the same time as the RCAF was assigned the task of answering this question demonstrates the interlocking roles of civil and military flying.

The RCAF's "aid to the civil power" was drastically reduced in 1932 owing to the same Depression-era economies that curtailed civilian airmail operations. When air force budgets were restored in 1936, they went largely to rearmament. However, "bush pilots in uniform" continued to operate as aerial map-makers. Immediately prior to the Second World War, they devoted much effort to charting the Labrador coast, where it was already anticipated that air bases and naval anchorages would soon be needed.

TECHNOLOGY ADVANCES

In February 1920, the Air Board established an Air Research Committee, the first of many bodies to study Canadian aerial problems, notably winter operations. Much work was delegated to universities. Within two years, studies had been published on aero engine performances at low temperatures (by the University of Alberta), antifreeze mixtures (McGill University), friction of skis on snow (McGill), cold-weather oil storage (University of Toronto), and effects of cold on aircraft rigging (Toronto).

Initially, airplane designs were ill-suited to Canadian conditions. Flying boats were seasonal machines, and open cockpit landplanes were extremely uncomfortable in winter. Enclosed cockpits did not enjoy wide usage until 1927, when Fairchild designs entered Canadian service, allowing the use of cabin heaters.

In frontier conditions, air-cooled engines proved more reliable than their water-cooled counterparts. At the end of the First World War, Britain's Armstrong Siddeley had developed the Lynx air-cooled engine. In the United States, the Wright company produced a very reliable air-cooled radial engine, the 220-horsepower J-4 Whirlwind, which achieved fame in 1927 when one powered Charles Lindbergh's *Spirit of St. Louis*

in the first non-stop solo crossing of the Atlantic. Pratt and Whitney, encouraged by American military authorities, designed the Wasp engine, which delivered over 400 horsepower. In Canada, radial engines were associated with Vickers Vedettes as well as Fairchild and Bellanca bush planes. As winter flying became more common, new problems came to light, including carburetor icing. This was met with directed engine heat and redesigned engine cowlings.

Extreme temperatures also affected oils and lubricants, which had to be thinned or drained on overnight stops. Operators devised nose hangars to shelter the engines and acetylene heaters to thaw them out in the morning. The solution to cold weather

A Mechanic's Solution

ON THE FIRST aerial foray into the far North in March 1921, two Junkers-Larsen JL-6 monoplanes belonging to Imperial Oil—christened *René* and *Vic*—became stranded at the Hudson's Bay Company post at Fort Simpson on the Mackenzie River. Accidents had resulted in smashed propellers on both machines, and the shipment of replacements would take months by the regular water route. Air engineer William Hill, assisted by a Hudson's Bay Company handyman, Walter Johnson, undertook to carve two propellers from oak sleigh boards that had originally been intended for dog sleds. Using one of the broken propellers as a pattern and employing Johnson's cabinet-making tools, they accomplished their task in eight days. To laminate the propellers they used glue made from moose parchments, which the Hudson's Bay Company had on hand. One of the newly made propellers was attached to *Vic:* it was perfectly balanced, with no trace of vibration. The skill, perseverance, and resourcefulness of Hill and Johnson had enabled them to produce a piece of highly technical equipment that has remained a monument to them both and an inspiration to future pioneers of the northern air trails. Both propellers are now part of the Canada Aviation Museum's collection.

lubricant problems lay in oil dilution—mixing gasoline with the engine oil, making the engines easier to start in cold conditions. The additives evaporated when engines heated up and were replaced after engine shut-down. The idea was articulated by Major Weldon Worth, of the U.S. Army Air Service, but developed to practicality by Thomas William Siers, maintenance superintendent for Canadian Airways. In the winter of 1940–41, that company reported engine cold starts at temperatures of −49°C (−57°F).

Wooden-hulled flying boats became waterlogged during the flying season, which substantially increased their weight. They were also vulnerable to many hazards when moving through water, such as submerged rocks and logs. Metal hulls or float-equipped aircraft, developed in the late 1920s, were not as susceptible to these hazards. In 1928, the RCAF and the National Research Council agreed to fund jointly a controlled-flow water channel that was used to test hull and float designs in the way wind tunnels assisted in aerodynamic research. Efficient floats became more than landing gear; extras such as canoes were sometimes strapped to the aircraft, and floats later doubled as water tanks for use in fighting forest fires during early post-war water-bombing trials.

Hinterland flying was adventurous, even dangerous. In the 1920s and 1930s, much of Canada was still unmapped. Pilots often used the "iron compass" (railway tracks) for visual navigation, but, beyond sight of the transcontinental lines, they had to make their own charts. Major rivers helped, but even these often fractured into mazes of indistinguishable lakes. While remote areas were being mapped, military and civilian pilots were assisted by the Northwest Territories and Yukon Signal System, developed by the Royal Canadian Corps of Signals. Their first wireless stations were established in 1923 at Dawson and Mayo, Yukon. Gradual expansion followed throughout the Mackenzie basin and beyond. By 1929, the Signal System's operations were directly harnessed to mineral exploration. Among the messages transmitted were the only comprehensive weather reports in the North. At some stations, a soldier acted as postmaster, airport superintendent, and justice of the peace as well as radio operator. The normal tour of duty was three years, but men bewitched by the North were known to request extensions. This communications

facing page: Historic artifacts in the Museum's collection include William Hill's hand-carved propellers that enabled the *René* and the *Vic,* the first aircraft into the far North, to be flown out following accidents that smashed their original propellers.

above: External loads could be strapped to a bush plane's floats, as on this Fairchild FC-2. Folklore has it that one enterprising pilot delivered an upright piano in this fashion.

above: As seen from this 1929 image, gauging ice thickness prior to landing with skis on frozen northern lakes and rivers was sometimes difficult, especially in late autumn or early spring when "rotten" ice was to be avoided.

facing page, top: A Fokker Universal being manhandled at Wakeham Bay, Québec, during the RCAF's Hudson Strait Expedition, 1927–28.

facing page, bottom: Practically minded mechanics and air engineers, collectively known as the "Black Gang" because of their perpetually grease-covered appearance, improvised stoves, nose hangars, and carburettor de-icers that kept engines running in the North's harsh climate.

network grew during the Second World War, and the construction of radar systems across the North in the 1950s completed the radio grid.

In partnership with the bush pilots were their mechanics and air engineers who serviced airplanes in bitter winter cold and summer insect swarms. Overhauling a floatplane or flying boat at its moorings required special care, lest tools or components be dropped into the water. Practically minded mechanics, often called the "black gang" because they were so often covered in engine grease, improvised the stoves, nose hangars, and carburetor de-icers that kept engines running.

The pilots themselves were a varied lot. Frederick J. Stevenson (1896–1928) was a rakish ex–fighter pilot, ex-barnstormer, and bush pilot, known to deliver forest fire fighters to a blaze and then join them in extinguishing the flames. Louis Bisson (1909–1997) aspired to the priesthood but chose instead to pilot Oblate missionaries about the North, pioneered transatlantic flying, and, in 1986, was finally ordained a priest. Between these two extremes were adventurers, businessmen, and dreamers, leaders in a future war (Zebulon Lewis Leigh) and at least two future airline tycoons (Grant McConachie and Maxwell Ward). In hard times they could sit idle for weeks, then be rushed into frantic dawn-to-dusk operations. Smoke from forest fires might blanket a region, stinging their eyes and complicating navigation. Moving freight and passengers entailed frequent landings and takeoffs in unfamiliar lakes, where depth, shoreline conditions, and underwater obstacles were unknown. Gauging ice thickness when landing with skis on frozen lakes and rivers was difficult and dangerous, especially in late autumn and early spring, when ice was at its thinnest.

Mechanical failure in remote areas occasionally led to pilots and passengers camping out for days until they were found or repairs made. Pilots most feared getting lost in the northern wilderness. When that happened, other pilots improvised search parties. In 1929, an extensive search failed to locate two aircraft and eight men of a prospecting party that went missing on September 8. The story ended happily on November 4, when the men walked into Cambridge Bay. A more successful operation was the 1936 search for Flight Lieutenant Sheldon Coleman and Leading Aircraftman Joseph Fortey, whose

CF-AWR

CANADIAN VICKE
G-CASK
Pratt and Whitney Engine Handbook
First Aerial Forest Patrol
IT IS THE DUTY OF EVERY CITIZEN TO HELP PROTECT OUR VALUABLE FORESTS
The St. Maurice Forest Protective Association
CANADA
AIR ENGINEER'S CERTIFICATE
The Guild of Air Pilots and Air Navigators of the British Empire.

aircraft was reported missing in the Northwest Territories; even so, a month elapsed between their forced landing and ultimate rescue.

Embodying the frontier spirit was the cache. To permit operations from distant bases, the RCAF and commercial firms used boats and aircraft to deposit fuel, lubricants, and canned goods at unmanned intermediary depots, the locations of which were—borrowing Inuit practice—made known to all. Anyone urgently needing these supplies could withdraw them from the cache, leaving a written record of what had been taken for future settlement of accounts. Everything was done on the honour system.

In 1921, new aircraft began to replace the First World War surplus machines. The Junkers-Larsen JL-6 led the way, with its seat belts and sheltered cockpit. In the United States, Sherman Fairchild designed aircraft to accommodate his new aerial cameras. His FC-2 aircraft of 1927 incorporated an enclosed cockpit. It was followed by the Wasp-engined Fairchild FC-2W and then by a succession of other Fairchild designs—the 51, 71, and Super 71—incorporating improvements suggested by frontier experience. Fairchilds were popular with civilian and military pilots as well as with managers trying to balance revenues with expenses.

Almost any airplane that could be fitted with skis or floats could become a bush plane. Over the years, their numbers included de Havilland DH.60 Moths, Curtiss Robins, Fokker Universals, a single-engine Junkers JU 52, Stinson Reliants, Cessna 180s, and twin-engine Beechcraft 18s. These were all adapted designs, but eventually the need for a purpose-built bush plane became obvious. The prototype Noorduyn Norseman flew in November 1935. Conventional in some respects, it introduced cabin insulation and wing flaps to bush flying. A large loading door accommodated fuel drums as well as cargo. During the Second World War, Norseman production was absorbed almost wholly by the American and Canadian air forces, but thereafter it became the quintessential Canadian bush airplane until other famous designs—the de Havilland Canada Beaver and Otter—overtook it.

Soon after bush flying began, a unique Canadian prize helped focus attention on aviation. In 1926 a wealthy American, James Dalzell McKee, wished to fly a Douglas seaplane across Canada and asked the RCAF for help. They provided Squadron Leader Albert E. Godfrey, a skilled, charming, and slightly eccentric officer, to serve as navigator. The flight proceeded in stages from Montréal to Vancouver, then down to California, with

facing page: Artifacts and memorabilia from the early days of bush flying. Shown is a leaflet dropped during the First Aerial Forest Patrol (lower centre), tools salvaged from the *La Vigilance* crash site (centre left), first day mail covers (centre right), and a blowpot heater used to pre-heat engines prior to start-up (upper centre).

facing page: The career of C.H. "Punch" Dickins (seen here in 1934), one of Canada's most skilled pioneer bush pilots, included a number of aviation firsts that helped unlock the secrets of Canada's north.

frequent stops at Ontario Provincial Air Service (OPAS) and RCAF bases. Impressed with the hospitality he had enjoyed, McKee created a trophy, formally called the Trans-Canada Trophy but more often called the McKee Trophy, to be awarded annually to the person making the greatest contribution to Canadian aviation in the previous year.

McKee was killed in an air accident before the first recipient was named. Initially, the trophy attracted little attention; early in 1928 only three persons were nominated. Fortunately, "Doc" Oaks was an outstanding candidate, and his receipt of the trophy gave it immediate prominence. The next two McKee Trophy winners were also distinguished bush pilots—Clennell Haggerston "Punch" Dickins, a pioneer of Barren Lands flying, and Wilfrid Reid "Wop" May, who had survived being chased by the Red Baron in 1918, and whose many mercy flights brought him rewards beyond that of the McKee Trophy.

THE TRADITION CONTINUES: POST-WAR FRONTIER FLYING

Commercial aviation was severely limited by the Second World War. Organizations that survived did so by keeping old machines airworthy while newer aircraft went to the armed forces or those civil operators most directly involved in war work. At the same time, Canadian companies sometimes signed transport contracts with the American forces for work along the Alaska Highway or in areas such as Newfoundland, where the United States had acquired bases. The transition from wartime to peacetime flying began in mid-1944—a year before the end of the war in Europe—when planning for post-war surveys and other tasks began.

In the immediate post-war years, hinterland flying picked up where it had left off. Surplus aircraft and demobilized aircrews were again involved in mineral exploration and transportation. One example was a former RCAF Consolidated Canso (CF-DTR), along with its crew captained by Squadron Leader John Hone. Leased by Frobisher Exploration Company, the aircraft left Mont Joli, Québec, on July 13, 1945, proceeding across Ungava towards Baffin Island. The men encountered floe ice, strong winds, dramatic wind shears, mountains, fog, and limited anchorages. They returned to Mont Joli on September 22, 1945, completing an expedition described as "ten weeks of successive brilliant exploits."

Some pre-war companies had disappeared, and new ones had been created soon after hostilities ceased. However, some bush firms operated across the decades. An example

C.H. "Punch" Dickins (1899–1995)

ONE OF the best known of Canada's early bush pilots, Punch Dickins dramatized the value of the bush plane in pushing aerial transportation into the far North. Many of his flights were pioneering efforts over desolate areas of the Northwest Territories that were designated "unexplored" on the maps of the day. Dickins had earned his pilot's wings in the Royal Flying Corps during the First World War and was decorated for gallantry under fire. At war's end, he returned to Canada and became one of the original officers of the Royal Canadian Air Force (RCAF) when it was formed in 1923.

He joined Western Canada Airways in 1927, and his career included an impressive number of Canadian aviation firsts that collectively helped unlock the secrets of Canada's north. The McKee Trophy winner for 1928, he was singled out for a 6 275-kilometre (3,900-mile) aerial survey expedition undertaken in a Fokker Super Universal, G-CASK, into the Northwest Territories and back to Winnipeg. He followed the western shore of Hudson Bay, and the highlight was a flight from Baker Lake southwest to Lake Athabasca that crossed the Barren Lands by air for the first time. Unlike earlier Arctic flights, which had followed known river routes, this was a true exploration of unknown, uncharted territory, from one watershed to another, with no certainty of finding suitable landing sites in the event of an emergency. Navigating over a confusing myriad of lakes, rivers, and muskeg, in unforgiving weather and out of radio contact with the rest of the world, Dickins often flew by the sun because his compass, affected by magnetic interference, couldn't be trusted. However, he did have faith in the new and dependable radial engines that were then becoming available through companies such as Wright or Pratt and Whitney.

In the summer of 1929, Dickins was the first to reach Canada's western Arctic coast by air when he landed at Aklavik, Northwest Territories, in a Fokker Super Universal of Western Canada Airways Limited.

For his outstanding aerial survey work and the development and expansion of flying routes in Canada's north, Dickins was named an Officer of the Order of the British Empire (OBE) in 1936. He was named an Officer of the Order of Canada in 1968 for his overall services to the nation. His legendary exploits as a pioneering bush pilot were followed by executive positions with Canadian Airways, Canadian Pacific Airlines, the Atlantic Ferry Service during the Second World War, and de Havilland Aircraft of Canada, where he developed a worldwide sales organization that sold Canadian-designed and -built aircraft to more than sixty countries.

was that of Arthur Fecteau of Senneterre, Québec. Gaining his private pilot's licence in 1931, he obtained limited commercial and air engineer certificates in 1936. Over the next thirty years, he graduated from a "bush-flying trapper" through to a one-man flying service and finally to the manager of a firm engaged in resource development, defence programs, and forestry protection. In 1949, his fleet consisted of four aircraft. By 1966, it had grown to eighteen aircraft. With one foot in bush flying's pioneer years and the other firmly planted in the modern era, Fecteau witnessed a revolution in frontier flying.

The return to pre-war working patterns was short-lived. Major changes had occurred between 1939 and 1945 that went beyond technical improvements such as advanced navigational aids. The Yukon and Northwest Territories had been sown with wartime airfields to assist in building the Alaska Highway, delivering lend-lease aircraft to the Soviet Union, and mining uranium for the first atomic bombs. Transatlantic ferry flying had brought similar construction throughout the northeastern Arctic, Newfoundland,

and Labrador. Although many airstrips were gravel outposts amid tundra and forest, they represented jumping-off points for ever more northerly surveys and development. In 1957, the most remote Arctic sites became accessible to aircraft with the introduction of large, low-pressure tires variously known as "tundra tires" and "Phipps specials." They had been developed and promoted by Welland Wilfred "Weldy" Phipps, who received the 1961 McKee Trophy for his pioneering work.

Airfields were gradually improved to accommodate more modern aircraft. On March 19, 1969, Nordair instituted scheduled jet services to the High Arctic using Boeing 737s. The machines landed at Hall Beach, 80 kilometres (50 miles) inside the Arctic Circle. Although the runway was still gravel, jets were equipped with "vortex dissipaters," which extended from under the leading edge of the engine inlets and directed three jets of high-pressure air downward to prevent the formation of whirlwinds that might pull debris into the engines. Large gravel deflectors were attached to the nose wheels and shielded the engines from gravel thrown up by the tires. These devices were modern aerial adaptations to frontier conditions.

If frontier flying became more respectable, it never completely lost its glamour. Incidents at the southern polar extremities demonstrate the portability of the bush pilot tradition. In January 1950, eleven men on a British research expedition were stranded on Stonington Island, southwest of Graham Land, Antarctica, due to ice and poor weather. A relief ship was turned back by pack ice. The British borrowed a Norseman and an RCAF pilot, Flying Officer Peter St. Louis, and moved them by sea to within 800 kilometres (500 miles) of the base. In two sorties, St. Louis evacuated five of the party. He also reconnoitred the ice conditions, enabling the rescue vessel to reach and retrieve remaining members. St. Louis's achievements would be echoed in April 2001 and September 2003 when Sean Loutitt of Calgary carried out dangerous medical evacuations from Antarctica in a de Havilland Canada Twin Otter.

facing page: Bob Cockram's Noorduyn Norseman, by Robert Bradford. The Norseman was the first purpose-built bush plane to be designed in Canada. It became the quintessential bush plane until another famous design, the de Havilland Canada Beaver, overtook it.

above: Violet Milstead and the Fairchild Husky she flew as a bush pilot for Nickel Belt Airways in the late 1940s. She became one of Canada's first female bush pilots following a wartime flying career with the Air Transport Auxiliary in the United Kingdom.

facing page: Roméo Vachon (seen here in 1928, fourth from left) was known as the "flying postmaster" for his work in linking remote communities along the north shore of the St. Lawrence River, often in adverse weather conditions.

In 1948, a *Canadian Aviation* magazine article ran the headline "The Lady Is a Bush Pilot." The subject was Violet Milstead, who had learned to fly in 1939. During the Second World War, she had joined Britain's Air Transport Auxiliary, shuttling aircraft around the United Kingdom. Returning to Canada, she took up instructing and married another pilot, Arnold Warren. They joined Nickel Belt Airways upon its formation in 1947. Soon she was piloting a Fairchild Husky, operating on wheels, skis, and floats, and having her share of forced landings and manual labour. Austin Airways bought the company in 1950. By then she had acquired an enviable reputation. One observer wrote:

> This pilot has steadily flown trips to all sorts of isolated places throughout the North and there is hardly a trapper, bushman, or prospector in this surrounding district that does not recognize this small parka-clad figure at the controls as she has at one time or other dropped them off or picked them up from some lonely lake far out and brought then back to civilization... The continuous and untiring performance throughout the past year from a base in Sudbury, Ontario, over rough terrain in all kinds of weather of this diminutive five-foot-four pilot is... worthy of praise...

Other newcomers to frontier flying have been First Nations personnel, who began appearing as commercial pilots in the late 1960s. Companies evolved under First Nations ownership, commencing with Air Inuit in 1979, followed by Air Creebec in 1982. Both firms now operate scheduled services as well as conducting charter work such as northern tourism and resource exploration. They have also attracted First Nations pilots; as of 2004, Air Inuit alone employed thirteen.

Bush flying speeded up Canadian access to the hinterland and far North. During recent times, companies such as Austin Airways, Kenn Borek Air Limited, and Bradley Air Services/First Air opened up many smaller, remote Northern communities to passenger and cargo traffic, operating such types as the de Havilland Canada Twin Otter and Hawker Siddeley HS748 turboprop aircraft. At the same time, the needs of frontier flying drove aeronautical development. The democracy of the frontier was ultimately demonstrated in the way flying began to change the demographic makeup of the North as the flying community was itself changed.

Roméo Vachon (1898–1954)

After wartime service as an engineer with the Royal Navy and enlistment in the Canadian Air Force in 1920, Roméo Vachon joined the Laurentide Pulp and Paper Company at the dawn of the bush-flying era. After qualifying as an air engineer he learned to fly, becoming one of Canada's earliest bush pilots flying Curtiss HS-2L flying boats on forestry patrol. He moved on to work for the Ontario Provincial Air Service (OPAS), which ultimately became the world's largest bush-flying operation.

In 1928, Canadian Trans-Continental Airways organized an air service under government contract to transport mail along the north shore of the St. Lawrence River, and they hired Vachon because of his familiarity with the area. While ferrying a new ski-equipped Fairchild monoplane, he parachuted a sack of mail onto the Québec City airport, the first time mail had been delivered in this fashion in Canada. Thereafter the "flying postmaster" routinely used this method of delivering the mail. During the next eleven winters, despite adverse weather conditions, his dream of uniting a string of isolated communities along the St. Lawrence through airmail service became a reality. He also encouraged the development of air strips, and radio and weather reporting stations, to improve the safety of all air transport operations in the area. Vachon was awarded the McKee Trophy in 1937, nominally for work the previous year but actually for his lengthy pioneering efforts in establishing airmail service in eastern Québec.

His subsequent aviation career included numerous outstanding achievements. In the late 1930s he became an executive with Trans-Canada Air Lines, and during the Second World War he organized the overhaul of aircraft for the British Commonwealth Air Training Plan. An award in his name was established in 1968 by the Canadian Aeronautics and Space Institute in memory of one of Canada's outstanding bush pilots. It is presented for the display of initiative, ingenuity, and practical skills relating to the art, science, and engineering of aeronautics and space in Canada.

the SECOND WORLD WAR 4

BETWEEN THE TWO world wars, the Royal Canadian Air Force (RCAF) waxed and waned, its very existence dependent upon its being useful rather than warlike. Its aircrews undertook aerial photography and mapping, carried survey crews to remote areas, flew missions to prevent smuggling and to enforce fisheries regulations, and transported government agents to remote First Nations reserves. None of this constituted preparation for a future war.

Still, the Canadian Air Force of 1920 (it acquired the "Royal" in 1923) spent much time training new pilots for current operations and future emergencies. Officers were routinely attached to Royal Air Force (RAF) units and attended British military schools such as the RAF Staff College and Imperial Defence College to absorb the latest air-power doctrines. In military exercises back in Canada, the RCAF practised directing militia and coastal artillery, scouting for troops, and dropping and picking up messages. In short, the RCAF was preparing to re-fight the First World War.

In the 1930s, as Europe and Asia drifted visibly towards another war, Canada began to rearm. Commencing in 1936, the RCAF began acquiring more current aircraft. Nevertheless, old theory persisted. The pre-war RCAF made no preparations for strategic bombing, and the Westland Lysanders that it obtained from 1939 onward were products of First World War concepts of aerial ground support. Slow and under-armed, the Lysander was

facing page: Robert Bradford's painting *DH Mosquito Night Fighter* depicts a successful night mission by a de Havilland Mosquito Mk. 30 of No. 406 Squadron, the top scoring RAF/RCAF intruder unit by war's end. Late in the war it was led by Wing Commander Russell Bannock, the RCAF's leading night fighter of the Second World War.

below: Personnel from No. 110 Squadron, RCAF, which was trained for army co-operation duties, departed in February 1940. They were Canada's first squadron to be sent overseas.

facing page: A damaged Messerschmitt Bf109F-1 being examined by Soviet troops. The Museum's Bf109F-4 suffered a similar fate in 1942, crash-landing in northern Russia after battle damage caused an engine failure.

vulnerable to both anti-aircraft fire and fighters. Its few virtues were ease of handling, manoeuvrability, and short-field performance. Despite these advantages, it showed that materially and intellectually, the RCAF was unprepared for the coming apocalypse.

The general European war that began in September 1939 so escalated that by 1941 only Ireland, Portugal, Spain, Sweden, and Switzerland remained neutral in the continent. In December 1941, Japan's entry broadened the war from India to Australia and throughout the Pacific Ocean. When the fighting ended in August 1945, more than fifty million people had died and the world had been transformed. The scope of the war is reflected in production figures. Never before or since have airplanes been built in such numbers. The United States alone turned out 303,000 machines. Great Britain produced 132,000, while Germany manufactured 120,000, Japan some 76,000, and the Soviet Union an estimated 158,000. Canada built around 16,000. The most numerous types of aircraft built were the Soviet Ilyushin IL-2 Shturmovik ground attack aircraft (36,200) and the German Messerschmitt Bf 109 fighter (28,000), very few of which survive.

The escalating capabilities of aircraft had a profound impact on the course of the war, and combat experience likewise affected aircraft evolution. Progress was most evident in matters of speed. The most modern piston-engine fighters of 1939 had a maximum speed of about 565 kilometres per hour (350 miles per hour); some piston-powered fighters in 1945 had a top speed of about 725 kilometres per hour (450 miles per hour). Fighter development was often a race between designers. In 1940, the Spitfire I (with its eight .303-calibre machine guns) and Messerschmitt Bf 109E were evenly matched. The Luftwaffe secured an advantage with the Messerschmitt Bf 109F; the

RAF regained some parity with the Spitfire V. Late in 1941, the Focke-Wulf FW 190 gave the Germans another advantage, and in 1942 the Spitfire IX countered with its 1,710-horsepower Rolls-Royce Merlin engine.

Propeller efficiency deteriorates as aircraft approach the speed of sound, but jet engines permitted greater speeds. The first jet, Germany's Heinkel HE 178, flew in August 1939 and was a test bed for the engine. Britain's Gloster E.28/39, flown in May 1941, was also experimental. Once jet engines were shown to be practical, designers hastened to apply them to combat aircraft. The high temperatures in such engines challenged metallurgical knowledge, however, and the earliest engines had very short running lives, so the introduction of jets into combat was long delayed. The first jet combat aircraft—Germany's Messerschmitt ME 262 and Britain's Gloster Meteor—became operational only in mid-1944: too late to affect the outcome of the war. Germany also introduced a rocket-propelled fighter, the Messerschmitt ME 163 Komet, but it was relatively unsuccessful.

Another, desperate experiment was the Heinkel HE 162, conceived in late 1944 to meet fuel shortages and Allied mass bomber formations. Designed for teenage Hitler Youth members, the Heinkel HE 162 Volksjäger ("People's Fighter") was to be mass-produced using wood not otherwise employed in aircraft construction. The scheme was a gamble bordering upon madness, for the HE 162 was inherently unsafe: it had fuel for

The Museum's Messerschmitt BF109F-4 is displayed with other significant aircraft of the Second World War. A Canadian-built Avro Lancaster X can be seen in the background.

only half an hour, and its control surfaces sometimes snapped off, sending the fighter out of control. The HE 162 was more dangerous to its teenage pilots than to any Allied aircrews.

Because of the difficulties encountered in being accepted into Canada's small air force, from the mid-1920s many Canadians travelled to Britain to join the Royal Air Force. Some 1,800 made this journey. Known as "CAN/RAF" personnel, some were in action from the very beginning of the war. Many participated in the 1940 Norwegian and French campaigns, before an RCAF presence had been established overseas, and before the war was over, 778 lost their lives. Ultimately, the numbers of CAN/RAF personnel would be dwarfed by the thousands of RCAF men and women arriving overseas.

The Canadian contribution to the war effort was enormous. As of August 1939, the RCAF had 4,061 men on strength. By December 1943, it had grown to 215,200, all ranks, including 15,000 enrolled in the Women's Division. They were assigned to three primary roles:

- The British Commonwealth Air Training Plan was a network of schools that trained personnel from throughout the Commonwealth in trades ranging from mechanic to pilot (see chapter 10).
- The Home War Establishment comprised bomber, fighter, and anti-submarine squadrons that defended both east and west coasts against possible attack, plus the transport and communications squadrons that wove the whole together.

The Museum's Heinkel He 162 Volksjäger was built near the end of the war using non-strategic materials such as wood. Intended to be flown by German youths with glider training, this jet fighter was not a machine for the inexperienced. Even seasoned pilots had to use delicate control movements to stay out of trouble.

› The RCAF overseas began with the movement of two squadrons to Britain in 1940. Later, more were formed or transferred overseas. Ultimately, forty-eight RCAF squadrons served in various roles and theatres. These squadrons often included substantial numbers of non-RCAF personnel and, at various times, at least half of all RCAF aircrew overseas served part of their tours in RAF squadrons.

Personnel varied as much as their trades. They included an estimated 8,864 American nationals who enlisted in the RCAF before December 7, 1941, when the United States entered the war. The best known of these was John G. Magee, whose poem "High Flight" is one of the most famous literary works associated with aviation. He was killed in a mid-air collision on December 11, 1941, one of approximately 800 Americans who died wearing RCAF uniforms.

At the outset of the war, the RCAF transferred almost its entire combat strength (which then comprised Westland Wapiti light bombers and Supermarine Stranraer flying boats) to Nova Scotia for anti-submarine patrols. Fortunately, no German U-boats appeared on the coast until November 1941, by which time the RCAF had acquired modern aircraft and weaponry to meet the threat. Eventually, six U-boats were destroyed by aircraft based in Canada.

In 1941–42, Canadian west coast squadrons were reinforced for fear of Japanese attacks. Curtiss Kittyhawk fighters and Bristol Bolingbroke bombers were deployed to Alaska during the Aleutian campaign of 1942–43, operating alongside American units. The only enemy aircraft destroyed by a Home War Establishment fighter was a Japanese Nakajima A6M2-N "Rufe" floatplane fighter, shot down over Kiska by a Kittyhawk of No. 111 Squadron (Squadron Leader Kenneth Boomer, pilot) on September 28, 1942. The Pacific Coast defences proved to be out of proportion to the threat, and they were reduced once Japanese forces had withdrawn from the Aleutian Islands in August 1943.

facing page, top: Hawker Hurricanes of the RCAF's No. 402 Squadron in England during what appears to be a staged scramble for publicity purposes, *c.* 1941.

facing page, bottom: Maintenance crew attending to a No. 411 Squadron Spitfire IX. This fighter squadron was the RCAF's fourth to be formed overseas during the war.

THE RCAF OVERSEAS

The RCAF entered combat in August 1940 when No. 1 (Canadian) Squadron participated in the Battle of Britain. Having won a defensive battle, RAF fighters gradually went over to the offensive, with many RCAF squadrons involved. On August 19, 1942, six RCAF fighter squadrons and two RCAF army co-operation squadrons were engaged over Dieppe. As of D-Day—June 6, 1944—there were ten RCAF Spitfire fighter squadrons in Britain. On December 29, 1944, Flight Lieutenant Richard J. Audet shot down five German fighters in one sortie, a feat unmatched by any other Spitfire pilot.

Only one RCAF overseas fighter squadron operated outside the European theatre—No. 417 in North Africa and Italy—but Canadian fighter pilots serving in RAF squadrons were ubiquitous. A CAN/RAF pilot, George "Buzz" Beurling, was the top-scoring pilot in the defence of Malta (twenty-six victories), while several RCAF pilots' careers also included time in Malta; Squadron Leader Henry W. McLeod, the most successful RCAF ace, gained twelve of his twenty-one victories there. Squadron Leader Robert W.R. Day destroyed five Japanese aircraft, becoming the only RCAF ace in the Far East. At least two RCAF fighter pilots served with RAF squadrons in the Soviet Union.

Another application of air power evolved under battle conditions: direct support of ground forces. Germany employed Junkers JU 87 dive-bombers and anti-tank aircraft to assist their armies. The Soviet Union used similarly specialized aircraft, notably the Ilyushin IL-2 Shturmovik. British and American air forces employed fighter aircraft in ground support work, using rockets, bombs, and gunfire to create havoc among opposing forces. Tactical air forces hindered enemy armies; in Normandy after the D-Day landings, German divisions dared not move openly by day. Three RCAF squadrons, grouped into No. 143 Wing, flew Hawker Typhoons from April 1944 to the war's end. The combination of low-level attacks and withering anti-aircraft fire entailed heavy casualties.

A decidedly "low-tech" form of tactical aviation also appeared. Early in the war, the Royal Artillery formed Air Observation Post (AOP) units to scout for armies and direct artillery fire, using such light aircraft as the Stinson 105 and Taylorcraft Auster. The American army followed suit, using the Piper Cub. Canadian troops became familiar with the AOP squadrons as they supported Canadian divisions in Italy and northwestern Europe. Three Canadian AOP squadrons were overseas in 1944 and 1945 with Auster IVs and Vs; only two of the units saw action, in the last month of the European campaign.

AE X
E S

The Evasion Kit

DOWNED ALLIED aircrew usually became prisoners of war, but thousands evaded capture with the help of escape kits and networks of brave Europeans who sheltered and fed them while moving them down "escape routes" at great personal risk. The escape kit included limited rations, maps, a compass, and European money. After D-Day, evaders in France and the Low Countries usually hid with civilians until advancing Allied armies reached them. A few evaders and escapees from prisoner-of-war camps went so far as to join resistance movements, sabotaging enemy communications and ambushing German troops. In so doing, they risked being summarily shot if captured.

In the 1920s and 1930s, theorists speculated that strategic bombing would determine the outcome of any future war. Bombers would carry death and destruction in the form of explosive and incendiary bombs and poison gas. (No one yet imagined nuclear weapons.) The 1936 film *Things to Come,* based on the H.G. Wells novel, showed Great Britain reduced to barbarism through bombing. Coupled with assumptions of total destruction was the belief in bomber invulnerability—that is, that the bombers would "always get through."

The development of radar in the mid-1930s modified these assumptions. As early as September 1939, German ground-based radar sets enabled their fighters to intercept and destroy British bombers in daylight, forcing the RAF to resort to night bombing. In 1940, British radar systems detected German bombers over France just after takeoff and directed RAF fighter interceptions so accurately that the Luftwaffe also abandoned daylight bombing and concentrated instead on nocturnal attacks. The two air forces, having been forced to adopt similar tactics, faced near-identical problems: for the attackers, how to navigate across blacked-out territory, find a target, and bomb it, and for the defenders, how to locate enemy night bombers and shoot them down in darkness. Again, the answer for both sides was radar.

Beginning with modified Bristol Blenheim aircraft, the night air war spawned specialized night fighters carrying radar, co-operating with ground stations and radar specialists who directed the fighters to the general vicinity of an enemy airplane. The early radar sets were unreliable and inaccurate, and the fighters were no faster than their intended prey. The Bristol Beaufighter, introduced in mid-1941, was the first effective Allied night fighter. Beaufighters were gradually replaced by faster de Havilland Mosquitos. Some night fighters, like the Messerschmitt Bf 110, were adaptations of earlier types, but an American aircraft, the Northrop P-61 Black Widow, was designed for no other purpose than nocturnal interceptions.

Airborne radar improved throughout the war, providing even more accurate range readings, greater definition, and enhanced reliability. Ultimately, however, the last few

facing page: Contents of an aircrew evasion kit, including a silk map and some currency, displayed near the tail turret of the Museum's Lancaster X.

below: The Museum's Mosquito B.XX was one of 1,033 "Mossies" manufactured by de Havilland Aircraft of Canada for the war effort, and served during 1944–45 at an Operational Training Unit in Nova Scotia.

facing page: In 1940, the RAF began operating four-engine bombers that carried triple the load of their predecessors. The Avro Lancaster was the most renowned British bomber of the war, and 422 Lancaster XS were built in Canada for the war effort.

seconds of any night encounter depended on keen eyesight and skillful co-ordination between the pilot and the radar observer.

RCAF night-fighter crews were everywhere. The first RCAF night-fighter squadron (No. 406) scored its initial victory on September 1, 1941. It was later joined by two other RCAF night-fighter squadrons, all operating Beaufighters and then Mosquitos with the most advanced types of radar for hunting in the dark. In addition, RCAF night-fighter crews served in RAF squadrons in northwestern Europe and the Mediterranean theatre. Intruder operations differed from classic, defensive night fighting in that intruders worked without airborne radar. They hunted enemy aircraft so far into hostile territory that authorities refused to issue them with the latest radar devices lest they fall into enemy hands.

Bombers evolved too. At the outbreak of the war, the Royal Air Force had two types of light bombers (the Fairey Battle and the Bristol Blenheim), each carrying about 450 kilograms (1,000 pounds) of bombs; the RAF also had three types of "heavy" bombers—the Handley Page Hampden, the Armstrong Whitworth Whitley, and the Vickers Wellington. These types could lift up to 1800 kilograms (4,000 pounds) of bombs.

As four-engine bombers began to be developed, these earlier "heavies" were reclassified as medium bombers. For sheer versatility, none surpassed the twin-engined de Havilland Mosquito, introduced in late 1940; its roles included those of day fighter, night fighter, photo-reconnaissance, anti-shipping, high-level bomber, low-level bomber, and even high-speed transport. Its lightweight wooden, laminated construction and twin Merlin engines made it one of the fastest aircraft of the war. A photo-reconnaissance version, the PRXVI, was the fastest version of the Mosquito, managing 680 kilometres per hour (422 miles per hour) in level flight. Since it could outrun almost any attacker, it had no need for armour. Mosquitos executed several precision attacks at low level, including destruction of Gestapo (German secret police) headquarters in The Hague and Copenhagen in April 1944 and March 1945 respectively.

Commencing in 1941, the RAF began operating four-engine bombers carrying triple the loads of their predecessors. First in action was the Short Stirling, followed by the Handley Page Halifax. A heavy twin-engine bomber, the Avro Manchester, was a failure owing to unreliable engines. But fitted with four Rolls-Royce Merlin engines, it was transformed into the Avro Lancaster, the most renowned British bomber of the war, associated with numerous famous raids including the breaching of hydroelectric dams in the Ruhr in May 1943 and the sinking of the battleship *Tirpitz* in November 1944. Lancasters also carried the biggest conventional bombs of the war, the 10 000-kilogram (22,000-pound) "Grand Slam." Meanwhile, the United States was building its four-engine bombers, the Boeing B-17 Flying Fortress and the Consolidated B-24 Liberator. American heavy-bomber design culminated in the Boeing B-29 Superfortress, which was used exclusively against Japan and ended the Pacific war with the atomic bombs dropped on Hiroshima and Nagasaki.

Allied heavy bombers waged a strategic air war intended to destroy enemy industry and demoralize the populace. Bomber Command's night-bombing campaign was difficult and dangerous. Special radio aids were developed for navigating across Europe. Devices were used to jam enemy radar; the Germans responded with their own anti-

Pilot Officer Andrew C. Mynarski, VC

DURING THE Second World War, thirty-four Victoria Crosses were awarded for aerial action. The last aerial Victoria Cross of the war was awarded to Pilot Officer Andrew C. Mynarski.

Mynarski was twenty-five years old when he enlisted and trained as an air gunner. He was ordinary in almost every respect; his post-war ambition was to return to the fur-cutting trade he had left in Winnipeg. Nevertheless, on the night of June 12, 1944, he performed a selfless act of heroism. As his aircraft was burning following a fighter attack, and with the other crew members already bailed out, he went through flames to attempt the rescue of the rear gunner, who was trapped in the turret. Mynarski struggled until his own clothes were burning. Giving up the effort, he returned to the escape hatch, saluted, said "Good night, sir," and jumped. He died shortly after reaching the ground.

The Lancaster crashed, but the gunner was thrown clear and miraculously survived. A year later, by sheer chance, gunner and pilot met in Canada during their demobilization. The story having been told, they wrote to the Royal Canadian Air Force headquarters, which suggested that "some award for Pilot Officer Mynarski" might be arranged. After months of investigation, authorities concluded that only the Victoria Cross was appropriate.

jamming inventions. The scientific duel became known as the "Wizard War." German night fighters hunted the bombers; British night fighters hunted the hunters. Enemy defences were formidable; in one night, March 30–31, 1944, Bomber Command lost ninety-six aircraft. By the end of the war, in May 1945, Bomber Command had lost 55,500 men in action or accidents and 9,838 had been taken prisoner. The U.S. Army Air Forces chose daylight raids, believing that such methods achieved greater accuracy. Over Europe, between August 1942 and April 1945, they lost some five thousand aircraft and 50,000 men (killed or captured).

facing page: One of the most renowned stories of Canadian heroism in Bomber Command was that of Pilot Officer Andrew Charles Mynarski, who gave his life trying to save a trapped crew member in their burning Lancaster. Portrait by Paul Goranson, 1947.

above: "Bombing up" a Handley Page Halifax of No. 6 Group, Bomber Command. The dolly carries containers packed with dozens of incendiary bombs that caused devastating fire damage at the target. Nearest is a single 2,000-lb (900-kg) high-explosive bomb.

Bombers were employed in ever-larger numbers. In May and June 1942, RAF Bomber Command made three raids using more than a thousand aircraft on each occasion. The newer "heavies" reduced the number of sorties flown while increasing the tonnage dropped. Thus, Wellingtons raiding Essen in 1941 carried approximately 1 800 kilograms (4,000 pounds) of bombs; a Halifax bomber raiding the same target in 1944 delivered about 4 300 kilograms (9,500 pounds) of bombs.

The first RCAF bomber squadron was formed in April 1941 and more soon followed. As of December 1942, there were ten RCAF heavy bomber squadrons overseas, and nine of these were brought together as No. 6 Group, the largest Canadian aerial formation ever assembled. The other squadron was incorporated into No. 8 (Pathfinder) Group. By mid-1944, No. 6 Group had grown to thirteen squadrons. Whether serving in RCAF or RAF bomber squadrons, Canadians fought against skilled enemies. Some 9,919 members of the force were killed as members of Bomber Command, more than half of all RCAF wartime fatalities.

MARITIME, RECONNAISSANCE, AND TRANSPORT WORK

Control of sea lanes was vital to both sides. German warships and then submarines strove to isolate Britain and the overseas war theatres from North America. Throughout the war, Allied aircraft hunted enemy ships and submarines, using increasingly

above: The long range of the Consolidated Liberator closed the "mid-Atlantic gap" in the battle against German U-boats. The Museum's example is one of thousands built by the Ford Motor Company in the United States. It is finished in the Second World War markings of Eastern Air Command, RCAF.

facing page: RCAF aircrew entering a Coastal Command Lockheed Hudson. This aircraft's crew consisted of a pilot, an observer, and two wireless air gunners.

specialized equipment. Some were land-based; others, like the Short Sunderland and Consolidated Catalina, were flying boats; the Canadian-built Canso (a version of the Catalina) was amphibious.

Nothing provides a better demonstration of submarine-hunting improvements than the Avro Anson and the Consolidated Liberator. In 1939, RAF Coastal Command used Ansons to scout for submarines. They had so little range that they could not cover the whole of the North Sea, and their 45-kilogram (100-pound) bombs were unable to sink a submarine. By 1943, Coastal Command had four-engine Liberators capable of coverage 1600 kilometres (1,000 miles) from base and equipped with advanced radar, powerful Leigh Lights to illuminate targets at night, and up to fourteen 225-kilogram (500-pound) depth charges that could destroy any submarine afloat. One-third of all German U-boats sunk during the war fell victim to aircraft or to ships operating in conjunction with aircraft.

Weapons continued to evolve with the aircraft. The 225-kilogram depth charge did not have to score a direct hit. It was fused to explode at a predetermined depth, and one blast close to a boat's hull could cripple or sink the vessel. Attacks required straight, low-level approaches, which exposed the aircraft to anti-aircraft fire if the enemy chose to fight on the surface. In 1943 the Allies introduced further refinements, including a torpedo that homed in on the propeller noise of a submerged U-boat and magnetic anomaly detection (MAD) devices that could find submerged U-boats. Both had only limited wartime success but would become primary anti-submarine weapons in the post-war era.

Germany, too, had its shipping to protect, notably iron-ore convoys that sailed from Norway. The Allies fought a relentless campaign against this maritime traffic. Prominent in this attack was the Bristol Beaufighter, which was armed first with a torpedo and then with rockets. Both weapons demanded a very close approach to enemy convoys, which were protected by ships bristling with anti-aircraft guns. Some fighters strafed the warships to suppress the flak while the main force attacked the convoy. Nevertheless, losses among the Beaufighter squadrons were very heavy.

Of eight RCAF squadrons represented in overseas maritime warfare, three were engaged in anti-shipping duties, although one later converted to anti-submarine work and another converted to heavy bomber duties. One maritime reconnaissance squadron

spent most of the war in Ceylon (now Sri Lanka). The RCAF overseas squadrons dedicated to anti-submarine work sank sixteen U-boats; aircrew in RAF units accounted for several more, including two in the Indian Ocean. The units engaged in attacking German shipping were credited with sinking sixty vessels (forty-one shared with other squadrons) and damaging many more.

Aerial reconnaissance was transformed during the war. Lysander aircraft were too vulnerable to survive in daylight warfare. In 1941, the RAF began using camera-equipped day fighters to reconnoitre the fringes of occupied Europe. The most ubiquitous of these were North American Mustangs and Supermarine Spitfires, working at low level and strafing targets if they had the opportunity (locomotives were favoured). Three RCAF squadrons operated in the tactical reconnaissance role; they were eventually consolidated into No. 39 Wing.

Changes in strategic reconnaissance had begun even earlier. Sidney Cotton, an expatriate Australian who had pioneered aviation in Newfoundland, proposed in 1939 that specialized high-speed, high-altitude aircraft might be the way of the future. Early in 1940, he was permitted to test his theories with a few borrowed Spitfires and pilots (three of them Canadians). The results were very successful. Increasingly, unarmed Spitfires and Mosquitos roamed over Europe, Africa, and Asia, photographing potential targets and assessing damage after air raids. The photo-reconnaissance (PR) aircraft scored several intelligence coups as well, including detection of German V1 and V2 developments—the precursors of cruise and ballistic missiles respectively. No Canadian squadrons operated in the strategic reconnaissance role, although many individual RCAF personnel flew in RAF PR units. The techniques of high-altitude photography covering large tracts of territory were subsequently applied to post-war aerial mapping back in Canada.

Aerial transport also revolutionized warfare. Germany used small numbers of paratroops in Norway, Belgium, and Denmark in 1940, and then mounted a costly aerial invasion of the isle of Crete in May 1941 using transport gliders and paratroops. The Allies learned from their defeats. In Europe, they mounted four large-scale aerial assaults to supplement conventional army invasions: Sicily in July 1943, Normandy in June 1944, the Netherlands in September 1944, and the Rhine Crossing in March 1945. Special forces were dropped into Burmese jungles and then provisioned from

facing page: Aerial transport revolutionized warfare. *Action at Swebo, Burma, January 12, 1945,* by Robert Bradford, depicts transport Dakotas of No. 435 Squadron RCAF, based in India, parachuting supplies to Allied troops operating under the nose of the enemy.

above: Aircrew trades were closed to women in the RCAF. However, a handful of Canadian female pilots joined the civilian Air Transport Auxiliary overseas to ferry aircraft from the factories to operational airfields.

facing page, top: With an endurance of almost 24 hours, the Consolidated PBY-5 Canso flying boat, shown here with beaching gear fitted, and its amphibious version, the PBY-5A, were used to counter the German submarine menace in the Atlantic and for patrols on the Pacific coast.

facing page, bottom: The action for which Flight Lieutenant David Hornell was awarded the Victoria Cross ended tragically for both Hornell and the U-boat he was stalking. In the hours following his attack on the submarine, Hornell would struggle to keep his men alive.

the air. An airlift from India over the Himalayas (known as "the Hump") supplied Chinese forces fighting Japan. Equally important were the thousands of transport sorties flown in all theatres that simply moved routine supplies about.

The RCAF was the first of the three Canadian services to enrol women for duties other than medical tasks. A Canadian Women's Auxiliary Air Force was authorized on July 2, 1941. Seven months later it was renamed the Royal Canadian Air Force (Women's Division); a member was routinely called a "WD." A specialized training depot opened in Toronto in October 1941, and graduates began reporting to RCAF units in January 1942. Ultimately, the air force enrolled 17,038 women (8 per cent of RCAF strength). They joined for diverse reasons, from adventurism to patriotism, but many enlisted because they had their own service connections, including brothers who had already signed up. Most served in Canada, but approximately 1,500 were posted overseas. The greater part of these served in RCAF Headquarters (London) or No. 6 Group Headquarters (Yorkshire).

Initially, the women were used in traditional "female" roles such as clerks, cooks, and telephone operators. Added trades came after April 1942: lab assistants, parachute riggers, band musicians—with mechanical and electrical trades following. At least thirty-five WDs were decorated for their services and twenty-eight members of the division lost their lives.

One trade was closed to RCAF women—that of aircrew. In 1940 and 1941, when some air force elementary flying training was being carried out by flying clubs, a handful of licensed women pilots secured instructional jobs. Those seeking more adventurous work looked abroad. Transatlantic ferry flying was also closed to them, but in Britain the Air Transport Auxiliary (ATA), a uniformed civilian organization, used women to ferry aircraft about the country, from factories to airfields and between bases. Five Canadian women joined the ATA, which gave them the opportunity to fly aircraft such as Supermarine Spitfire fighters, Vickers Wellington bombers, Douglas Dakota transports, and Fairey Barracuda torpedo-bombers. Two more Canadians joined a similar American organization, the Women's Airforce Service Pilots (WASPs).

Flight Lieutenant David Hornell, VC

THE CANSO is forever enshrined in Canadian aviation history through its association with Flight Lieutenant David Hornell, VC. On June 24, 1944, flying Canso "P" of No. 162 Squadron, RCAF, he and his crew attacked and sank *U-1225* in the North Atlantic but were shot down by the submarine's anti-aircraft guns. Hornell ditched successfully and all his crew escaped from the airplane, but nothing went well afterwards. One dinghy was lost; the crew of nine had to take turns sitting in the remaining raft or holding on to its side in the water. Through a twenty-one-hour ordeal, Hornell offered inspired leadership and spent more than his share of time in the water. Two of the crew died of exposure and Hornell succumbed himself minutes after being rescued by marine craft. For the attack on the U-boat and his subsequent example in the water, he was posthumously awarded the Victoria Cross.

facing page, top: Personnel from the RCAF's Women's Division engage in semaphore training (visual signalling with flags) at a wireless school in Montréal.

facing page, bottom: Initially, RCAF female personnel were employed in traditional roles such as office support, but training in mechanical and electrical trades soon followed.

Precisely how many members of the RCAF served abroad is unknown, but if—as likely estimates have it—the force recruited about 250,000 (all trades and ranks), perhaps 150,000 went overseas at one time or another. As of 1939, the RCAF had about 500 aircrew, and the British Commonwealth Air Training Plan (BCATP) trained a further 72,835, making a total of approximately 73,335 aircrew. Again, we do not know precisely how many served abroad, but 60,000 might be a high estimate. We know the casualty figures with greater certainty. For home and abroad, they were as follows:

	FLYING	NON-FLYING	TOTAL
OVERSEAS	14,332	338	14,670
HOME	1,787	671	2,458
TOTAL	16,119	1,009	17,128

Even with plans for a Canadian force in the Pacific, the RCAF was undergoing reductions before Victory in Europe (VE) Day. (Between December 1944 and May 1945 it declined from 193,000 to 164,500 personnel.) Victory in Europe and Asia brought rapid demobilization, and by December 1945 it was down to 58,000. The last RCAF squadrons in Europe—part of the Occupation Force in Germany—disbanded in March 1946; five units in Britain disbanded shortly afterwards, and in June 1946 the three transport squadrons were sent to Canada, taking their Dakotas with them.

In the rush to peace, an important symbolic event occurred. In 1922, the Canadian Air Force had adopted the Royal Air Force ensign as its own. In June 1940, the King had approved an RCAF ensign, which included a Maple Leaf within the RAF roundel. It was flown throughout Canada and wherever Canadian squadrons were based overseas. But the RCAF had continued to display the RAF roundel on its aircraft, whether in Canada or abroad. On January 17, 1946, the RCAF adopted a new roundel for its aircraft, the version that appeared on the 1940 RCAF ensign. It was a typically Canadian gesture—confidently done, quietly expressed—confirming the nation's pride and satisfaction in more than six years of struggle.

OM

naval AVIATION

5

THE VALUE OF naval aircraft had been recognized even before the First World War, but developing the machines and methods took many years. In January 1911, Eugene Ely, an American flyer, landed a Curtiss biplane on the afterdeck of the battleship USS *Pennsylvania* and took off again—though the ship was anchored in San Francisco Bay at the time. The following year, Charles R. Samson (Royal Navy) took off from a warship steaming at sea. In August 1917, also at sea, Squadron Commander E.H. Dunning (Royal Navy) landed a Sopwith Pup on a short, improvised flight deck fitted to HMS *Furious.* He had to dodge the ship's funnel and forecastle before alighting, a manoeuvre so dangerous that Dunning crashed on his second attempt two days later and was killed. Meanwhile, the Royal Navy launched occasional air strikes from warships using land planes flying one-way missions. Thus on July 19, 1918, seven Sopwith Camels took off from *Furious* to bomb Zeppelin sheds in Germany. Three Camels returned to alight in the sea (they were lost; their pilots were rescued), three landed in neutral Denmark, and one vanished without a trace.

The naval version of the Sopwith Camel used in this raid (the 2F.1) had a shorter wingspan than the land-based F.1, a fuselage that could be separated into two sections for shipboard storage, and a different gun configuration. During this time of experimentation, Camels were launched from platforms constructed over a ship's gun turret

facing page: This collection of naval artifacts includes the leather flying helmet (centre) that once belonged to "Mel" Alexander, the last surviving member of Raymond Collishaw's famous "Black Flight" of Naval Squadron 10 (see page 34), an all-Canadian unit of First World War fighter pilots. Bottom left is a standard-issue Webley service revolver.

or took off from rafts ("lighters") towed by destroyers. Whether intercepting aircraft or Zeppelins or bombing inland targets, the naval Camels were sacrificed after completing their missions.

AIRCRAFT CARRIERS EVOLVE

The most important and fundamental problem to be solved before the first successful deck landing carrier could be built was the breakup of airflow over the deck caused by the ship's superstructure, mast, and funnel.

The first British aircraft carrier with a flush deck (ship's superstructure entirely removed, providing an unencumbered landing platform) was a converted ex-Italian passenger liner, the HMS *Argus* of 1918, which carried fifteen to twenty aircraft. HMS *Hermes*, the first British ship to be designed from the start as an aircraft carrier, was commissioned in 1923 and had an offset "island" incorporating the command centre and funnel; it also carried fifteen to twenty aircraft. In 1922 the USS *Langley*, the U.S. Navy's first carrier, added arrester gear and crash barriers to snag aircraft before they ran off the deck. Development of these and subsequent vessels produced many innovations, including sealed storage of aviation fuel, elevators (lifts) to move aircraft between hangar and flight decks, and aircraft with folding wings to facilitate storage.

The Sopwith Camel has become the most widely recognized British fighter of the First World War. Like the Museum's example, this 2F.1 naval version is fitted with a different gun configuration from the more common land-based F.1 Camel.

The Washington Naval Treaty of 1922 put international limits on battleship construction. Consequently, several capital ships then under construction were redesigned as aircraft carriers, which at the time were considered defensive and so were counted under a separate category in the treaty. Naval officers and strategists debated how these vessels would be used. All navies assumed that battleships were still at the heart of sea battles. In the 1930s, Japan and the United States grappled with the concept of task forces using two or more carriers. Carrier proponents were not encouraged when German surface ships sank HMS *Glorious* with all her aircraft aboard in June 1940. Nevertheless, aircraft carriers achieved a victory in November 1940, when a Royal Navy air strike crippled three Italian battleships in Taranto harbour. At the Battle of Cape Matapan in March

1941, Fleet Air Arm torpedo-bombers damaged Italian warships at sea. In May 1941, the German battleship *Bismarck* was crippled in mid-Atlantic by carrier-based Fairey Swordfish torpedo-bombers; the vessel's destruction was completed by pursuing British warships.

A Royal Naval Air Service officer was the first airman to land on a moving vessel when Squadron-Commander E.H. Dunning skidded his Sopwith Pup onto an improvised flight deck fitted to HMS *Furious* in August 1917. Note the holding straps on the aircraft's lower wing.

Japan's aerial attack on Pearl Harbor on December 7, 1941, demonstrated again what carrier-borne aircraft could do to a fleet in harbour. So far, however, no such aircraft had ever dispatched, unassisted, a major enemy vessel at sea. On December 10, 1941, just days after Pearl Harbor, land-based Japanese bombers and torpedo aircraft sank the British battlecruiser HMS *Repulse* and battleship HMS *Prince of Wales* in the Indian Ocean, the first capital ships to be sunk solely by air power on the open sea. In April 1942, Japanese carrier-based planes sank the British carrier *Hermes*, which had no aircraft aboard for protection, and the Australian destroyer HMAS *Vampire* in the Indian Ocean, demonstrating Japanese superiority in carrier operations. Early in May 1942, American and Japanese forces fought the Battle of the Coral Sea, the first naval battle waged entirely by opposing carrier-borne aircraft. On June 4 and 5, 1942, another all-air battle near Midway Island changed the entire Pacific war; one American and four Japanese carriers were sunk. Again, the opposing ships met only by proxy, through their fighters, torpedo-bombers, and dive-bombers.

The Pacific theatre was marked by large, dramatic, and costly carrier campaigns in which the U.S. Navy (joined in 1945 by the Royal Navy) swept the Japanese navy from the seas and supported a series of island-hopping campaigns. Elsewhere, large aircraft carriers played other roles. They delivered fighters to Malta in the Mediterranean, escorted convoys to the Soviet Union, supported amphibious landings from Madagascar to Morocco to Italy, harried German shipping, and attacked (but failed to sink) the battleship *Tirpitz* in harbour. When that ship (the final capital ship of the Axis powers active in the European theatre) was finally knocked out by Royal Air Force (RAF) Avro Lancasters, Britain dispatched carriers to the Pacific, attacking targets that included oil refineries, ships, and airfields.

On a smaller scale, convoys needed protection against long-range German bombers. Several merchant ships were hastily equipped with catapults to launch Hawker Hurricane fighters, which would be ditched after a mission. (At least three Canadians became "Hurri-cat" pilots.) A more useful solution was the conversion of merchant ships (often former grain carriers or tankers) into small carriers with three to five aircraft. These Merchant Aircraft Carriers, better known as "MAC ships" or "baby flat-tops," hunted German submarines throughout the Atlantic with considerable success.

An aircraft often launched from MAC ships was the Fairey Swordfish. Nicknamed the "Stringbag" because it could be loaded up like an old-fashioned grocery bag, the Swordfish torpedo-bomber looked old-fashioned, yet it remained in service and production longer than its intended successor, the Fairey Albacore. It entered Fleet Air Arm service in 1936 and achieved fame at Taranto, at Cape Matapan, and in the pursuit of the *Bismarck*.

Canada was introduced to the Swordfish through the MAC ships. A Royal Navy (RN) depot at Halifax (HMS *Seaborn*) assembled and serviced Swordfish for the carriers, and in 1942 No. 1 Naval Air Gunners School was established at Yarmouth, Nova Scotia, using Swordfish and training over seven hundred gunners. At the end of the war, most "Stringbags" were scrapped, but twenty-two were turned over to the Royal Canadian Navy (RCN), which used them for training and communications flying until 1948.

facing page: Because the correct identity of the Museum's Fairey Swordfish is unknown, it was assigned a fictitious number during its restoration that represents an aircraft known to have been on Royal Canadian Navy strength from 1946 to 1948.

CANADIAN NAVAL AVIATION

Canada watched carrier warfare with interest. A few Canadians had joined Britain's Fleet Air Arm (FAA) before the Second World War, but direct Royal Canadian Navy (RCN) participation in naval aerial operations came more slowly. Late in 1940, a group of 150 members of the RCN were sent to England to train in specialized fields for an expanding RCN. Eighteen chose aircrew training. Having won their wings, they joined Royal Navy FAA units, although they still wore "Canada" flashes on their uniforms. They saw action both ashore and aboard British carriers from Norway to the Pacific Ocean; six were killed in action or flying accidents, including Lieutenant Robert Hampton Gray.

Early in 1943, the RCN proposed formation of a Canadian naval air branch, starting with systematic naval aircrew training. Some of the graduates were in action before the war's end, and like Gray, flying from British carriers. Meanwhile, the Royal Navy, short

above: The escort carriers HMS *Puncher* and HMS *Nabob* were manned with Canadians but employed British aircrews. A type initially flown from their decks was the Fairey Barracuda, the Royal Navy's first monoplane torpedo bomber, seen here *c.* 1944.

facing page: Lieutenant Robert Hampton Gray, VC (1917–1945), was involved in attacks on the *Tirpitz* and died after helping sink the *Amakusa.*

of ships and men, welcomed any Canadian participation in naval aviation. In 1943, two small escort carriers, HMS *Puncher* and HMS *Nabob,* were set aside to be Canadian-manned ships. (The aircrews would be British, flying Grumman Wildcat fighters as well as Fairey Barracuda and Grumman Avenger torpedo-bombers.) The *Nabob* subsequently had a short but active career off Norway, its aircraft harrying German shipping and the *Tirpitz* itself, but on August 22, 1944, it was torpedoed by a submarine. Although the *Nabob* managed to limp back to port, its war was over. The *Puncher,* which had been commissioned five months after the *Nabob,* saw action off Norway in 1945. Following Victory in Europe (VE) Day it became a training vessel, then a troopship, before being transferred to the United States in January 1946.

Soon after the war's end, Canada had a light fleet carrier, HMCS *Warrior,* originally built in Belfast for the RN and later loaned to the RCN. The naval aircrews would be Canadians who either had served with the RN or had been trained since 1943 for a distinct Canadian naval air branch. A number of former Royal Canadian Air Force (RCAF) aircrew also volunteered for the new air arm. Former Fleet Air Arm squadrons were disbanded in Scotland late in 1945, their British personnel scattered, and the units then reformed as RCN units with as many Canadians as could be scraped together. They were then brought to Canada to be incorporated into the post-war RCN.

Working in Britain, the first Canadian naval air squadrons flew Barracudas, Fairey Fireflies, and Supermarine Seafires. Happily, the Barracudas never crossed the Atlantic; they were disliked by many in the Fleet Air Arm and were described by one writer as a "monument to misapplied ingenuity." The Fireflies were better; initially used as long-range reconnaissance aircraft, they were adapted to anti-submarine work by the fitting of advanced radar plus sonobuoys. When dropped into the ocean, these could detect submarine noises and broadcast them. Seafires were naval versions of the Spitfire, differing from the "Spit" in having folding wings, arrester hooks, and strengthened undercarriages. The *Warrior* remained with the RCN until 1948, when it was replaced by HMCS *Magnificent* ("Maggie"), another borrowed light fleet carrier.

Robert Hampton Gray, VC

ROBERT "HAMMY" Gray of Nelson, British Columbia, joined the Royal Canadian Navy Volunteer Reserve in July 1940. He volunteered for Fleet Air Arm training in December of that year, graduating as a pilot in September 1941. In August 1944, he joined No. 1841 Squadron aboard HMS *Formidable* as a Chance Vought Corsair fighter pilot. He was subsequently engaged in attacks on the *Tirpitz* as well as in air actions with the British Pacific Fleet. These experiences earned him a Mention in Dispatches and a Distinguished Service Cross.

On August 9, 1945, he led a formation at low level to bomb Japanese warships in Onagawa Bay, including a 1000-tonne warship, the *Amakusa.* They met fierce anti-aircraft fire from ships and shore batteries. Gray's aircraft was set on fire and one of his bombs was shot off the wing, but he pressed home his attack. His remaining bomb hit and sank the *Amakusa.* Just after passing over the target, his machine burst into flames, rolled to the right, and crashed into the sea. He was posthumously awarded the Victoria Cross, having displayed "great valour... a brilliant fighting spirit and most inspiring leadership."

below: Fairey Fireflys preparing to launch in about 1947, with Hawker Sea Furys in the background. The Fireflys were long-range reconnaissance and fighter aircraft later adapted to anti-submarine work and were one of the types flown by the first Canadian naval air squadrons aboard HMCS *Warrior* and HMCS *Magnificent*.

facing page: A Grumman Avenger AS Mk. 3 launches from HMCS *Magnificent*. Sturdy Avengers, modified for anti-submarine operations, began to replace the RCN's Fairey Fireflys in 1950.

While the RCN had been planning its naval air arm, the Pacific war was still in progress and the navy had hoped to operate two carriers. The Japanese surrender made that unnecessary, and the post-war naval establishment of ten thousand men rendered a two-carrier plan impractical. In January 1947, the government reduced the RCN even more; its budget was cut by $50 million and the staff ceiling dropped to 7,500. By 1948, the fleet had been reduced to forty-four ships. The naval air arm centred on the sole carrier, but the force also used major land bases, including Royal Canadian Naval Air Service (RCNAS) Shearwater, at the mouth of Halifax harbour, which until 1948 had been RCAF Station Dartmouth. RCN aircraft and crews also exercised regularly at the Canadian Joint Air Training Centre in Rivers, Manitoba. The flight training of RCN aircrews differed little from that of their RCAF counterparts.

The immediate post-war RCN looked very different from what it had been during the Second World War and from what it became when committed to the North Atlantic Treaty Organization (NATO): a specialized anti-submarine force. The flotillas that sailed in 1947–48 would have been appropriate to the operations anticipated off Japan had the Pacific war continued.

Naval forces conducting fleet exercises at sea were seldom seen by civilian audiences. The coronation of Elizabeth II in 1953 gave the RCN an opportunity to show off its Sea Furies and Avengers in a Commonwealth aerial pageant at Spithead, England. On other occasions, the *Magnificent* was more of a transport than a carrier; in 1951, it moved No. 410 Squadron, RCAF, to Britain with forty-eight Canadair Sabres lashed to the deck. In December 1956, it sailed from Halifax, its flight deck jammed with vehicles destined for the first United Nations Emergency Force in Egypt, and in February 1957 the deck was again used to transport old Sabres from Europe to Canada.

Ship technology was matched by aircraft evolution. The *Warrior* had embarked Seafires

and Fireflies, but in 1948 the Seafires gave way to Hawker Sea Furies. The Sea Fury was a carrier version of the Fury, which had itself evolved from the Tempest and first flew in September 1944. The Fury was probably the finest piston-engine fighter developed in Britain, with a top speed of 780 kilometres per hour (485 miles per hour). With jet aircraft appearing, however, the Royal Air Force lost interest, but the RN ordered it as a carrier aircraft. It saw action in Korea with British and Australian carriers. The RCN acquired its first of seventy-four Sea Furies in 1947 and flew them until 1956.

Replacement of the Fireflies began in 1950 with sturdy Avengers, most of them modified for anti-submarine operations but eight equipped with Airborne Early Warning (AEW) radar. The RCN now trained its air arm to hunt and kill submarines; the small force was so specialized that it could not participate in the Korean War, where no submarine threat existed and carriers were used chiefly to launch tactical air strikes on ground targets.

HMCS *Bonaventure* permitted the operation of even heavier aircraft. The types adopted were the McDonnell F2H-3 Banshee and the Grumman CS2F Tracker, the latter

HMCS Bonaventure

In 1957, the Royal Canadian Navy exchanged the *Magnificent* for HMCS *Bonaventure* (promptly dubbed "Bonnie"), which was owned (not leased) by Canada. The ship had been built with the latest carrier technology. Its angled deck (with flying conducted some 25 degrees away from the bow) divided the flight deck into separate parking and runway areas. This design actually enlarged the flight area and eliminated the need for crash barriers (though not arrester hooks). A steam catapult enabled the launching of heavy jet aircraft in moderate winds and even when stationary. The new Mirror Landing Sight, which combined visual and audio features, simplified landings by projecting a stable light beam from deck to aircraft; pilots could then fly down the beam without reference to a "batman" waving corrections during approaches. Throughout its career, the *Bonaventure* conducted 20,590 deck landings.

manufactured by de Havilland Canada under licence. Unhappily, the Banshees strained even "Bonnie" and could operate only in the most favourable weather. The ship normally handled eight Banshees and twelve Trackers, but as early as May 1958 senior officers complained that neither figure constituted a viable force, either as an anti-submarine unit or as a defensive fighter flight. The Banshees were more significant through their assignment to the North American Air Defense Command (NORAD), where their fast rate of climb and Sidewinder missiles made them formidable interceptors. They were retired in 1962, even before the *Bonaventure* itself departed, in controversial circumstances.

The increasing complexity of anti-submarine warfare (ASW) meant that the Tracker required a crew of four (pilot, co-pilot, and two naval aircrew ASW sensor operators) compared to three crew members in the Avenger. Radar and visual spotting ("eyeball Mark I") spotted submarines on the surface. Submerged submarines were detected by sonobuoys (the Tracker carried sixteen) and a magnetic anomaly detection (MAD) device in the tail.

By the mid-1960s, the RCN was having second thoughts about a fixed-wing naval air arm when land-based anti-submarine aircraft, such as the Canadair Argus, had extended mid-Atlantic reach. Shortly after the announcement of Canadian Armed Forces unification (which eventually took place in 1968), the directorate responsible for naval flying went through five reorganizations in one year. In this confusing atmosphere, the *Bonaventure* went into dry dock for a mid-life refit (1966–67). The estimated cost had been $8 million, but the final bill was $17 million. This led to attacks on the concept of an RCN air arm, culminating in "Bonnie" being decommissioned on July 3, 1970, and scrapped. Thereafter, the only seaborne air presence in the Canadian forces were helicopters operated from ships. The Trackers continued to serve from land bases in such roles as sovereignty patrols and fisheries surveillance until 1990, when they were finally retired.

facing page, top: Its 2,480 horsepower Bristol Centaurus radial engine made the Hawker Sea Fury one of the most powerful piston-engine fighters ever manufactured, and the last to serve in the RCN. The Museum's aircraft was one of 74 that served aboard HMCS *Magnificent* and on shore bases.

facing page, bottom: After HMCS *Bonaventure* was scrapped, the only aircraft to be operated from Canadian Forces ships were helicopters such as this Sikorsky CHSS-2 Sea King, about to land on the destroyer HMCS *Assiniboine* in 1963. The ship is equipped with a hangar, landing platform, and a haul-down system (at the centre of the flight deck).

NAVY 110
CANADIAN
NAVY

234
NAVY

403

HELICOPTERS

6

THE IDEA OF a flying machine capable of hovering flight and vertical takeoff is very old indeed. Leonardo da Vinci sketched designs in the early 1480s, and by the nineteenth century the basics of aerodynamics were understood. However, development of machines capable of true vertical control did not occur until well into the twentieth century.

The first helicopter flights took place in France in 1907, but these early flights were virtually uncontrolled. Among the challenges of early hovering flight were the demands of power (sufficient to lift a craft by the rotors alone), control (how to keep the craft from rotating in the direction opposite to its own rotor), and progression (how to fly forward rather than merely hover). The first two problems were usually attacked using two rotors turning in opposite directions; the third was for a time insoluble.

The Spaniard Juan de la Cierva (1895–1936) fitted freewheeling rotor blades on a pylon attached to an Avro 504 airframe, producing an autogyro. The first successful flight was accomplished in 1923 and by 1928 he had developed his machine to the point of making a flight across the English Channel. In succeeding years, he granted development licences that allowed British, German, American, and Soviet companies to explore autogyro technology. In 1930 Harold Pitcairn produced the first North American version at his factory in Willow Grove, Pennsylvania: the three-seat, 300-horsepower PCA-2.

Unlike that of a helicopter, the autogyro's rotor is not powered; it turns in response to the machine's forward motion (produced by a standard, powered, forward-facing

facing page: A Sikorsky Sea King engages in a "vert-rep" operation, naval speak for "vertical replenishment" of ship's stores. Taken on strength by the RCN in 1963, Sea Kings were still in service with the Canadian Forces more than forty years later.

The second version of the VS-300, piloted by Igor Sikorsky. The assistant's sign reads "WORLD'S RECORD BROKEN 1 HOUR 20 MIN" for hovering flight.

propeller) and the resultant action of the air on the rotor's blades. By turning, the rotor provides lift, doing away with the need for wings. In the case of the earliest autogyros, someone on the ground had to spin the rotor before the pilot began to taxi. However advanced the models, the aircraft had to move forward quickly to take off. Thus, although autogyros could take off and land within short distances, they were incapable of true vertical or hovering flight.

As the autogyro reached its peak, helicopter evolution accelerated. In 1935, Louis Breguet, who had been experimenting with helicopters since 1907, flew a successful machine that attained speeds of 110 kilometres per hour (67 miles per hour), altitudes of 170 metres (560 feet), and flights of 62 minutes. Germany's dual-rotor Focke Achgelis FA 61 of 1936 introduced the method of controlling directional flight by varying rotor blade pitch while airborne. It was dramatically displayed hovering inside a sports stadium.

Igor Sikorsky (1889–1972), who had experimented unsuccessfully with helicopters in Russia (1909), emigrated to the United States, where he became a successful flying-boat designer. He returned to helicopter experiments with his VS-300, which first flew on September 14, 1939. This experimental machine was modified several times until, by 1941, it had been configured with a single main rotor to achieve vertical lift and a small tail rotor to counter the torque and offer directional control. Sikorsky also worked out the principles of the flexible rotor head that were to become the standard method of control for all helicopters that followed. The flexible rotor head employed a sliding collar on the rotor shaft called the "swashplate," which was linked to two control levers in the cockpit. This tilting and sliding swashplate pushed up or pulled down on rods linked to the rotor blades, allowing their pitch to be changed either individually (cyclic pitch) or all together (collective pitch). The collective pitch lever enabled the machine to take off vertically, hover, and land vertically. By manipulating the cyclic pitch lever, the main rotor could be tilted forward, backwards, or sideways, allowing the machine to move in those directions.

While European progress stagnated during the Second World War, American development, led by Sikorsky, was dramatic. His XR-4, test flown in January 1942, employed an

Although they could take off and land within short distances, autogyros were incapable of true vertical or hovering flight. This Pitcairn PCA-2 was flown in the Trans-Canada Air Pageant of 1931.

engine of 185 horsepower, more than double the power of the VS-300. More significant was the dependability of the type, which was ordered into production as a military trainer but was soon being used in pioneering operations in the field. The Sikorsky S-51 appeared in February 1946, followed by the S-55, test flown in November 1949. Both types were manufactured in the United States and under licence in Britain, and they were widely exported.

The Bell Aircraft Corporation began experimental work on helicopters in 1943 and test flew its Model 47 in 1945. It soon became the world's first commercially successful helicopter. Carrying a pilot and one passenger, it was adaptable to cargo and casualty evacuation; above all, the Bell 47 was small enough to get into and out of restricted spaces, including ships' decks and oil-rig platforms. Built under licence in several countries, it remained in production until 1974, by which time 6,439 had been built.

In the Soviet Union in 1946, Mikhail Mil was appointed head of a Rotating Wing Scientific Research Laboratory. He had been engaged in autogyro work since 1931, and he soon produced the Mil-1, an outstanding utility helicopter that remained in production until 1965, with 3,500 being built in the Soviet Union and Poland. By 1952, Soviet designers had caught up with their Western counterparts, and Russia continues to produce excellent machines, civil and military, to this day.

9609
OU 609
ARMY

Early helicopters powered by piston engines had limited lifting powers. Turbine engines offered greater power-to-weight ratios. The first turbine-powered helicopter, the French Sud Aviation Alouette II, flew in 1955. It was quickly purchased by military and civilian customers, Canadians included. The availability of more flexible machines brought a rush of orders for a bewildering array of helicopters from Britain, France, the United States, and the Soviet Union. By the late 1970s, Canada was importing about a hundred helicopters per year.

facing page, top: In 1946, the Bell 47 became the world's first type-certified civil helicopter. The Model 47G, shown here in Canadian Army service, remained in production until 1974.

facing page, bottom: Aside from a few naval officers previously instructed in the United States, Canada's military was introduced to helicopter flight in the late 1940s after the RCAF's acquisition of its first helicopter type, the Sikorsky H-5 (S-51).

HELICOPTERS IN MILITARY AND NAVAL SERVICE

Helicopter history includes many military milestones. The first combat medical evacuation involved a Sikorsky R-4 flown by a Royal Air Force pilot in Burma in April 1944. British use of helicopters in Malaya, starting in May 1950, was overshadowed by a much larger deployment of "choppers" during the Korean War. The first use of helicopters in assault roles (that is, delivery of troops directly to a battlefield and use of air-to-ground weapons) was by the British in Malaya and the French in Indo-China. The techniques and technology were subsequently expanded by all major armed forces, to the point that helicopters have become familiar actors in news film footage from Vietnam, the Falklands, Afghanistan, the Persian Gulf, and Iraq.

Canada's first helicopter pilots were naval officers sent to the United States for instruction in 1944. The Royal Canadian Air Force (RCAF) acquired seven Sikorsky S-51 helicopters in 1947, designating them by the name H-5; they were used principally to train pilots and to familiarize personnel of all three services with such craft. In 1951, the Royal Canadian Navy formed a helicopter flight at Shearwater with Bell HTL-4s. The next year Sikorsky HO4S helicopters (Helicopter Observation, or "HO4S" in naval terms, "Horse" to those who worked with them) were introduced into service for communications and "plane guard" (rescue at sea) duties. Experiments with helicopters in the anti-submarine role began in 1955 using dipping sonar. These were successful, but adapting "choppers" to ships other than aircraft carriers was more difficult; the HO4S was too light and fragile for sustained operations in the restless mid-Atlantic.

In the late 1950s, the Royal Canadian Navy (RCN) began searching for a suitable machine that could operate from the new generation of destroyer escorts. This led, in

The aircraft nomenclature system introduced by the U.S. Army in the 1950s included the letters HU ("helicopter utility"). From this the unofficial nickname "Huey" was derived. Serving in the Canadian Forces, the Bell CH-135 "Twin Huey," seen here, was powered by a pair of Canadian-built turbines.

1963, to the RCN's adoption of the Sikorsky HSS-2 Sea King helicopter, a type still in service forty years later with both the Royal Navy and Canadian Forces. Throughout their service history the Sea Kings have been operated in gruelling conditions. Those deployed to the Persian Gulf in 1991 as part of the Canadian contribution to liberating Kuwait were stripped of their anti-submarine gear and modified for several perceived threats, including Iraqi missiles and chemical weapons. Once on station, the Canadian ships and their Sea King helicopters challenged vessels suspected of running a blockade imposed on the area and helped small craft out of dangerous waters.

Initially, the Canadian army left helicopters to the RCAF and RCN. In 1961, the army acquired light Hiller CH-112s for observation and liaison flying with troops in Europe. Three years later, they added Vertol CH-113 Voyageur helicopters for heavy lifting. The Hillers were withdrawn from service in 1972 and replaced with Bell CH-135 Twin Hueys and Bell CH-136 Kiowas. To this fleet was added the Bell CH-146 Griffon in 1996.

The Huey gained a high profile during news coverage of the Vietnam War. The model had first flown in 1956, and the U.S. Army designation HU-1 ("Helicopter, Utility") gave rise to its nickname, "Huey." Bell later adapted the airframe to a paired engine layout using a Canadian-designed engine, the Pratt and Whitney Canada PT-6T Twin Pack. This inspired the nickname "Twin Huey." The Canadian Armed Forces acquired fifty of these, which were designated CH-135s.

NON-COMBAT AND CIVIL APPLICATIONS

In 1946, Canadian mineral exploration companies used a leased American Bell 47 helicopter near Sudbury, Ontario. The following year, Allan Soutar registered the first Canadian commercial helicopter (Bell 47 CF-FJA) on behalf of the Canadian Photographic

Survey Corporation in Oshawa, Ontario. Two months later, Carl Agar of Okanagan Air Service brought in another Bell 47 (CF-FZX).

These pioneers devoted a lot of time to demonstrating the potential uses of "choppers." Soutar showed off the crop-dusting capabilities of his machine to Ontario farmers and other uses to the Ontario Provincial Air Service.

The most dramatic pioneer work was performed by Carl Agar (1901–1968), a former RCAF instructor who had formed Okanagan Air Service immediately after leaving the forces. Hearing of helicopters working in the state of Washington, Agar drove south to investigate and returned to Canada converted to "choppers." While the performance of

Early use of the Bell 47 proved the viability of helicopter operations in the most inhospitable conditions and locations. This machine is possibly CF-FZX, Carl Agar's first helicopter, converted to include a bubble canopy, *c.* 1954. Agar's company, Okanagan Air Services, soon became known as the world specialist in mountain helicopter operations.

A Shearwater Angel

THE ROYAL Canadian Navy's HO4S helicopters carried special equipment for rescue work, including a hoist capable of lifting 270 kilograms (600 pounds), a hoisting sling, and a bullhorn speaker to communicate with those being picked up. The Canada Aviation Museum's machine, 55877, has a dramatic history. It often took its turn as the "Shearwater Angel," the standby search-and-rescue helicopter at RCNAS Shearwater, and markings on the nose tell part of the story—eight rescues over fifteen years, including five carried out at sea.

Its most daring mission, on November 26, 1955, resulted in the award of George Medals for valour to pilot Lieutenant-Commander J.H. Beeman and co-pilot Lieutenant-Commander F.R. Fink; two crewmen were commended for Brave Conduct. Both types of awards are indicated on the helicopter. The rescue involved the retrieval of twenty-one seamen from the freighter *Kismet II,* which had grounded on the coast of Cape Breton Island. Turbulence, heavy seas, and the location of the ship at the base of a 120-metre (400-foot) cliff compounded the difficulties of the task and the heroism of the helicopter crew.

helicopters was still uncertain, Agar began testing their abilities at mountain altitudes. On one occasion, he alighted on top of a 2 300-metre (7,500-foot) cliff. He managed to lift off again into the thin air, barely hovering using ground effect, then dropped into a 600-metre (2,000-foot) abyss, plummeting 150 metres (500 feet) before he had enough forward speed for his rotors to take effect.

Whether packing supplies for topographic survey crews, surveying timber stands, airlifting material for construction of a dam, or moving prospectors, Okanagan Helicopters (Agar's company name from 1949) proved the viability of helicopters in the most inhospitable conditions. Agar was awarded the 1950 McKee Trophy, but that was not the end of his pioneering. He had graduated from Bell 47 to Sikorsky S-55 machines and used them in the High Arctic for cargo transportation. In 1987, Okanagan Helicopters merged with several other companies to form the CHC Helicopter Corporation. It eventually absorbed Australian, British, Canadian, and Norwegian firms, demonstrating the global reach of such companies.

If Agar proved that helicopters could operate almost anywhere, Robert Heaslip (1919–) showed they could carry almost anything. Unlike Agar, he stayed with the RCAF after the Second World War and trained as a helicopter pilot in 1947. In June 1954, he formed No. 108 Communications Flight, the first all-helicopter unit in the RCAF. This coincided with the construction of the Mid-Canada Line radar detection chain. The flight operated six Sikorsky H-34 helicopters, six Piasecki H-21s, and ten Sikorsky H-19s, which between them hauled 10 000 tonnes of material and fourteen thousand people to various construction sites from Labrador to British Columbia. Loads were both carried internally and slung externally, winter operating procedures were devised, and pilots and crews were pushed to their limits. Heaslip was awarded the McKee Trophy in 1956 in recognition of his pioneering contribution.

Early use of helicopters in rescue missions could not disguise the fact that the machines of the time were not suitable for all such tasks. The Sikorsky R-4, which figured so prominently in such missions, had a very short range and could evacuate only one person at a time. However, helicopter development was rapid; soon the Sikorsky S-51 and S-55 brought more power and lifting capacity, features that were amply demonstrated during the Korean War.

facing page: Navy HO4S-3s (Sikorsky S-55s) aboard HMCS *Bonaventure*. These helicopters were widely used by the Royal Canadian Navy for rescue work, and one hovered continuously on guard during carrier flying exercises at sea.

In the mid-1950s, RCAF helicopters were assigned to specialist search and rescue units. Piasecki H-21s stood guard for more than a decade, saving lives in often harrowing conditions.

In 1955, RCAF helicopters were assigned to specialist search-and-rescue units. Even so, performance was sometimes wanting in hot weather and in the thin air at high elevations; in July 1957, an S-51 evacuation was requested from mountains near Canmore, Alberta, but could not be conducted because the victim was 1800 metres (6,000 feet) above sea level. Technical progress removed this handicap, as was demonstrated in April 1965 when a Boeing Vertol CH-113 Voyageur landed on the summit of Mount Hubbard, Yukon, 4 557 metres (14,950 feet) above sea level.

In 1975 the army's CH-113 Voyageur operations changed to air force jurisdiction, and the former army machines were renamed CH-113 Labradors, in keeping with those already operated by the air force. In its vital search-and-rescue (SAR) role for the air force, this twin-engine, twin-rotor helicopter was a workhorse for over forty years until the last of the fleet were retired in 2004. It incorporated a watertight hull for water landings, a side-mounted rescue hoist, and a 5 000-kilogram (11,000-pound) cargo hook. The Labrador's 1,500 shaft-horsepower General Electric T-58-GE-8F gas turbine engines gave it a cruising speed of over 200 kilometres per hour (125 miles per hour) and it had a range of 1 110 kilometres (690 miles). Some of Canada's most intense SAR missions have involved Labradors. In 1997 alone, SAR teams responded to more than seven thousand calls for help.

Plucking survivors from imminent danger has highlighted "chopper" capabilities, but aid from the sky is often a matter of prolonged operations. During the Manitoba Red River floods of 1997, and following the massive 1998 ice storm in Ontario and Québec, civilian and military helicopters flew thousands of hours, reconnoitring the damage, sometimes retrieving stranded civilians, but more often transporting key equipment and personnel to remote or inaccessible locations.

Helicopter applications have evolved with global priorities and societal demands. Just as mineral exploration and mining spurred frontier aviation between the world wars, so has the global thirst for oil encouraged helicopter sales since the 1960s. On land, from Alaska to the Middle East, helicopters have been used to facilitate exploration and

CH-113 Labrador #11301 on the day of its delivery to the Museum in 2004, the last Canadian Forces Labrador to fly. As it was the first of its type to be acquired by the RCAF in 1964, it was especially appropriate that it also make the last flight. Its replacement, the EH Industries CH-149 Cormorant, can be seen in the background.

A Sikorsky S-61 belonging to Cougar Helicopters of St. John's, Newfoundland, on its oil rig landing pad. This company became the primary helicopter operator in the Hibernia field.

development. At sea, they have been vital links to offshore rigs. Geography has been no barrier; in 1964, Okanagan Helicopters and British European Airways created International Helicopters to service North Sea drilling rigs.

In Canada, systematic development of Newfoundland's Hibernia field, 335 kilometres (180 miles) southeast of St. John's, began in 1991. Cougar Helicopters of St. John's became the primary operator, using Eurocopter Super Puma machines equipped with the most modern Global Positioning System of receivers, de-icing equipment, and high-intensity lighting to avoid "missed approaches" at fog-shrouded rigs. Only pilots with a minimum of three thousand hours of flying time were hired. Given the cold water temperatures (even in summer), it was mandatory for all passengers and crew to take survival training and to wear immersion suits when flying between land and rigs. Cougar Helicopters also maintained a constant rescue capacity, including medical evacuations at one hour's notice. The loss of eighty-four lives in February 1982, when the Grand Banks drilling rig *Ocean Ranger* was overcome by storms, has never been far from the minds of oil companies and aviators.

Urban applications have included police work, medical evacuation, and radio reporting of traffic conditions. Helicopters have been widely involved in tourism projects. The stately Ford Trimotors, which offered aerial views of Niagara Falls from 1928 onward, have been succeeded by nimble choppers swooping close to the mist. Niagara Helicopters Limited began such flights in 1961. Another tourist application has been helicopter skiing expeditions in the Rocky and Coast Mountains. (Carl Agar would be surprised and pleased.)

As early as 1943, New York City had investigated police applications for helicopters. Some twenty years passed before American agencies thoroughly studied helicopters in law enforcement, notably in California, at Long Beach, Lakewood, and Los Angeles. Canadian police forces leased helicopters sporadically until 1971, when the Royal

Canadian Mounted Police (RCMP) acquired their first rotary-wing aircraft, a Bell 212. Even so, such machines were more often used for transport than for active police enforcement. In 1977, the Durham Regional Police Force, east of Toronto, established and integrated a helicopter unit with constables on the ground. Numerous cities have since formed similar units. Opinions differ about the degree to which helicopter patrols deter crime, but there is no debate about their usefulness in surveillance of urban disturbances and in car chases.

Although helicopters were initially identified with wilderness or high-seas rescues, they eventually found a place in the urban medical system. Their tasks have ranged from neonatal emergencies to rapid movement of organs for transplant. Since 1970, most new major hospitals have been designed to include a helicopter landing pad, and many older institutions have witnessed helicopters landing and taking off from adjacent lawns, highways, or parking lots.

Since helicopters can operate from almost any large, stable platform, they have become ubiquitous—from the decks of icebreakers and Coast Guard vessels to the rooftops of corporate headquarters. Various companies offer a staggering array of services: environmental protection, forest-fire suppression, geological survey, logging, mail delivery, medical evacuation, mining, pipeline construction, political campaigning, security, silviculture, tourism, and transmission-line monitoring, to name but a few. The greatest measure of helicopter growth is found in statistics: In 1947, there were five commercial helicopters registered in Canada; in 2003, there were approximately thirteen hundred.

Dini Petty, "The Pink Lady"

COMMERCIAL HELICOPTER companies long resisted hiring female pilots, arguing that isolated bush regions were "no place for a lady." But cities were a different matter. Competing with other radio stations that were reporting traffic conditions using helicopters, CKEY (Toronto) decided to employ a female pilot. They hired Dini Petty, paid for her helicopter flying lessons, and in 1968 launched her career as Canada's second commercially licensed, female helicopter pilot. (Marion Orr, a former Air Transport Auxiliary pilot, was the first.)

Flying a pink Hughes 300 helicopter and wearing a pink flying suit, Petty was a popular celebrity, whose work involved solo flying, spotting, and reporting. An instant hit with the Toronto public, her celebrity status grew and reporters in the United States billed her as the first woman in the world to fly a helicopter and report on traffic.

Petty later said that flying was terrifying at first, because of the stress of trying to co-ordinate flying the aircraft, being in contact with the control tower and the other CKEY helicopter, and listening to the radio station so she wouldn't miss her cues. A bad day might include a scratchy VHF radio, a blaring AM receiver, a complicated ground traffic situation, and a sick passenger. Regarding herself primarily as a broadcaster who had become a helicopter pilot, she left after four years and four thousand hours in the air to pursue a successful radio, television, and writing career.

447
402

7 PEACE *and the* COLD WAR

CANADA BEGAN TO prepare for peace long before the end of the Second World War. In June 1944, air force recruiting ceased. As long as the Pacific war continued (and a role for Canada was envisaged), there could be no final decisions for the post-war armed forces, but after Victory over Japan (VJ) Day, demobilization proceeded apace, accompanied by reorganization of what remained. Royal Canadian Air Force (RCAF) strength fell from 164,850 in May 1945 to 58,050 by December 1945 and then to 12,735 by December 1946. The Women's Division was disbanded completely; women would not be recruited again until May 1951.

facing page: View from the "office" of a McDonnell CF-101 Voodoo. Voodoos served in the Canadian Forces until the 1980s, when McDonnell Douglas CF-18s joined the northern watch.

The world, however, proved to be more dangerous and threatening than had been expected in 1945. Fighting in Europe had scarcely ceased when the Soviet Union and its former allies quarrelled over how defeated Germany and other eastern European territories should be governed. Many hoped that diplomacy would solve these differences, but a *coup d'état* in Czechoslovakia and a Soviet blockade of the U.S., British, and French sectors of Berlin in 1948 set Western Europe, the United States, and Canada on a path leading to rearmament. In 1949, a defensive alliance was formed: the North Atlantic Treaty Organization (NATO). The Czech coup could not be reversed, but the Berlin blockade was broken in a massive operation known as the Berlin airlift (1948–49).

The International Civil Aviation Organization (ICAO), based in Montréal, was formed in 1944 to encourage the orderly growth of civil aviation, which had quickly assumed

Wartime developments in aerial reconnaissance led to photographic techniques used later in mapping Canada's north. This B-25 Mitchell bomber, seen in 1944, mounted three cameras that took both side-angled and vertical shots for wide photographic coverage of the terrain below.

global proportions. One agreement was that countries should assume responsibility for search-and-rescue operations within their airspace and adjacent sea zones. In 1946, Canada designated the RCAF to be responsible for such operations. This was a radical change. Previously, air searches had been organized and conducted on an ad hoc basis. Now the process was organized, with designated units and search centres that could marshal civil and military resources to seek lost aircraft or assist crash survivors.

Wartime developments in aerial reconnaissance led directly to a revision of aerial mapping techniques. Throughout the world, various air forces employed high-altitude aircraft with multiple, high-resolution cameras. Simultaneously, specialized photo labs and cartographers applied the new methods.

In Canada, much of the mapping was "north of 60," the great swath of the northern part of the country, including the Arctic Archipelago. Most impressively, 2.4 million square kilometres (911,000 square miles) were photographed and 12 950 square kilometres (5,000 square miles) of new territory were discovered as Arctic islands in 1948. No. 408 Squadron, based at Rockcliffe and flying Avro Lancaster bombers converted to photographic roles, were at the heart of the program.

JET FIGHTERS AT HOME AND ABROAD

Jet engines suck air into combustion chambers where kerosene-based fuels burn at high temperatures. This air is compressed by either axial-flow or centrifugal-flow compressors at the front of the engine and is mixed with the fuel in the combustion chambers. The resulting exhaust gases are then expelled from the rear of the engine, creating thrust. Commencing in 1928, a Royal Air Force cadet, Frank Whittle (1907–1996), pioneered many design concepts, and his engines were flight tested in 1941. Technological sharing of Whittle designs enabled the United States to test its first jet airplane, the Bell XP-59A Airacomet, in October 1942. When German technology became available after the war, it was studied, compared, and adapted to new designs.

The de Havilland Vampire Mk. 3 was Canada's first operational jet fighter. Its forward fuselage section was constructed of wood and accommodated a pressurized cockpit and four 20-mm cannon.

The world's air forces entered the jet age with fighter aircraft that had been designed around these first engines but which were, in most other respects, fairly conventional machines. The Gloster Meteor, de Havilland Vampire, Bell P-59 Airacomet, Lockheed P-80 Shooting Star, and Mikoyan Gurevich MiG-9 were all remarkably similar in performance, and they all began to make headlines. The world air speed record, which stood at 755 kilometres per hour (470 miles per hour) in 1939, was raised by a Meteor to 975 kilometres per hour (605 miles per hour) in 1945, while a Vampire executed the first jet landing and takeoff from an aircraft carrier.

Canadian experience with early jet fighters was not uniformly successful. The de Havilland Vampires acquired in 1947 were delightful machines to fly, but their Perspex canopies often frosted over and sometimes cracked. Spare parts were hard to find, which influenced Canadian thinking against buying British aircraft. Instead, Canada focussed on American- and Canadian-built machines.

The second generation of jets followed quickly. They exploited research by German engineers—particularly into swept wings—and experiments in the immediate post-war period. Whereas civilian aviation in the 1920s and 1930s had often been ahead of military technology, in the late 1940s the military led the way, so that the Boeing B-47 Stratojet bomber, with swept and flexible wings, pointed the way to the Boeing 707 and succeeding airliners.

above: Canadian-built F-86 Sabres were the first swept-wing aircraft to serve with NATO forces in Europe. The RCAF squadrons based in Britain, France, and Germany trained constantly, ready to repel invaders.

facing page: Cover of sales brochure for the Rolls-Royce Nene, 1950.

The RCAF of 1947–48 was unprepared for operations abroad, as it was preoccupied with mapping, flood relief, and Arctic supply missions. Canada did not participate in the Berlin airlift, although at least three Canadians did, two as members of the Royal Air Force (RAF) and one RCAF pilot on exchange duties with the United States Air Force (USAF). Expansion and rearmament accelerated in 1948–49, but when the Korean War broke out in June 1950, Canada found itself in a dilemma: Should it throw its air power into the new Asian war or concentrate on European defences? There was no simple answer.

The first Canadian commitment to the Korean War was three navy destroyers; the second was No. 426 Squadron and its Canadair North Stars, which were assigned to a trans-Pacific airlift. A Canadian army brigade followed, but the RCAF's principal focus was to create and maintain the fighter defences of Britain and Western Europe. Starting in 1951, a succession of Canadair Sabre squadrons was formed in Canada and transferred to Britain, France, and Germany. By 1955 No. 1 Air Division was fully deployed in Europe, ready to fend off a Soviet invasion that never came. Even so, the RCAF deemed it wise to have some pilots with jet combat experience, so between 1950 and 1953, twenty-two Sabre pilots were attached to USAF squadrons in Korea. On March 30, 1951, Flight Lieutenant J.A.O. Levesque became the first Canadian to shoot down an enemy aircraft in all-jet combat. The most successful RCAF Sabre pilot in Korea, Flight Lieutenant E.A. Glover, destroyed three MiG-15s.

Prototype F-86 Sabres and MiG-15s, which both had swept wings to reduce buffeting as they approached the speed of sound, first flew in 1947. The Soviet MiG-15 could climb higher and faster than the Sabre but suffered instability at high speeds. Designed to shoot down bombers, the MiG-15 was armed with 23-mm and 37-mm cannons that tore large holes in its target but fired too slowly for effective fighter-versus-fighter combat. The Sabre, with rapid-firing .50-calibre machine guns, scored heavily against MiGs.

The Sabre squadrons in Europe trained constantly, ready to repel invaders. The Canadians were respected by other NATO pilots and three times won the Guynemer Trophy as the most accurate air-to-air marksmen in Western Europe. The Guynemer competitions and the Korean War dogfights represented a passing era, however, for missiles, radar, and laser guidance systems would soon supersede combat in which men paired off against one another in machines equipped with guns. But success and respect came at a price: 107 Sabre pilots lost their lives in NATO service.

As Cold War policies changed, four of the Sabre squadrons in Europe were replaced by Avro Canada CF-100 all-weather squadrons. Then, between 1961 and 1965, the Air Division was reduced from twelve to eight squadrons and the Sabres were replaced with

The Goblin and Nene

WHITTLE ENGINE patterns were developed rapidly into more powerful power plants that were also more economical in fuel consumption. The de Havilland Goblin of 1943 provided 1400 kilograms (3,100 pounds) of thrust to power the earliest Vampire and P-80 fighters. In 1944, Rolls-Royce designed the Nene engine, which delivered some 2300 kilograms (5,100 pounds) of thrust. The Nene powered a new generation of aircraft, including the Canadair T-33 (developed from the P-80), the Grumman F9F Panther, and the Hawker Sea Hawk. Nenes that were exported to the Soviet Union before Cold War tension rose to a critical level became the basis of the Klimov VK-1 engine that powered the Mikoyan Gurevich MiG-15.

facing page: The Lockheed F-104 Starfighter replaced the North American F-86 Sabre as the RCAF's NATO fighter in Europe. Both were American designs manufactured in Canada by Canadair, and powered by engines built by Orenda. CF-104s equipped eight squadrons overseas in the dangerous low-level strike role.

Canadair CF-104 Starfighters, equipped for high-speed, low-level reconnaissance and air strikes. Had bombs been used, they would have been nuclear weapons, and the realities of atomic warfare meant that if war had broken out, most pilots would have flown only one mission before both sides were reduced to radioactive rubble. Whether in the reconnaissance or strike role, Starfighter pilots routinely practised at high speeds and low altitudes, trusting their ejection seats if trouble occurred. In all, eighty-four Starfighter pilots ejected safely following incidents that ranged from bird strikes through engine failures to mid-air collisions; a further thirty-seven were killed, including seven whose parachutes failed to deploy properly following ejection.

Canada sought a non-nuclear air force from its NATO commitments. By 1972, the CF-104s (reduced to four squadrons) had assumed conventional roles. The Air Division had become an Air Group in 1970, and when it was re-equipped with McDonnell Douglas CF-18s (1984–86) it was reduced to three squadrons, although each new aircraft type was capable of feats unimaginable to its predecessors. Although the RCAF (later, the Canadian Forces) had prepared to fight a war for forty years, it did not fire a shot in anger until 1991, when CF-18s participated in the first Persian Gulf War. In 1999, it took part in NATO operations aimed at halting ethnic cleansing in Kosovo. In that campaign, Canadian CF-18 pilots flew 678 sorties, roughly 10 per cent of all NATO missions.

The defence of Europe was matched by concern over the defence of North America. Since 1942 Canada and the United States have co-ordinated their activities; the process accelerated after 1947, culminating in a formal agreement in 1958 that created NORAD (North American Air Defense Command). Initially, the fear was that the Soviet Union, having tested its first atomic bomb in 1949, might attack North America with bombers, using Arctic airspace as a shortcut. The threat was exaggerated, the response consequently enormous. In the next decade, three lines of radar stations were built from coast to coast: the Pinetree Line (early 1950s), the Mid-Canada Line (mid-1950s), and the Distant Early Warning (DEW) Line (late 1950s). These passive defences were backed by three thousand interceptor fighters (two hundred of them Canadian) and ninety Nike and Bomarc missile squadrons (two of them Canadian). They were directed from computerized regional control centres known as SAGE (Semi-Automatic Ground Environment) sites dug into Cheyenne Mountain in Colorado Springs and into the Canadian Shield in North Bay, Ontario.

704

From 1953 to 1964, Canadian fighter squadrons deployed in North America were equipped with a Canadian design, the CF-100—known within the RCAF as "The Clunk"—a dependable machine that eventually served in nine home-based units. Its origins lay in the Canadian authorities' 1945 decision that henceforth national airspace would be defended by aircraft suited to Canadian geography. Since no suitable airplane existed, Avro Canada designed the CF-100 to RCAF specifications. The prototype, with Rolls-Royce Avon engines, flew in January 1950; subsequent aircraft (Mark 2 to Mark 5) had Orenda engines developed and built in Canada. Changes included enlarged radar sets (most evident in the bulbous nose of the Mark 4) and evolution from an all-gun armament (Mark 3) to an all-rocket armament system (Mark 5).

Historians do not agree on when the era of long-range bombers ended and that of intercontinental missiles began, but cancellation of the costly Avro CF-105 Arrow interceptor program in 1959 signalled that bombers were now deemed less threatening. The CF-100s were replaced instead by two squadrons of Boeing Bomarc missiles and five squadrons of McDonnell CF-101 Voodoos, all acquired from the United States. The Voodoos served until the 1980s, when CF-18s joined the northern watch. The CF-18 was intended to replace three earlier aircraft (CF-101, Canadair CF-5, and CF-104) and to operate in both European and North American environments. Unlike the earlier Sabres and Starfighters, the CF-18 was built in the United States, but with some features unique to Canadian service. The most striking was the addition of a 600,000-candlepower spotlight on the left forward fuselage for night identification of other aircraft.

Throughout the Cold War, RCAF/Canadian Forces aircraft intercepted Soviet ones only sporadically, usually Soviet airliners that had strayed off course en route to Cuba or reconnaissance aircraft testing NORAD alertness. No bomber fleets arrived. The collapse of the Soviet Union and the end of the Cold War in 1989–90 has since raised questions about the continued maintenance of manned fighters waiting for a non-existent enemy.

facing page: Pilots were impressed by the versatility, comfort, instrumentation, and computers of the McDonnell Douglas F/A-18 Hornet. Designated CF-188 in Canada, the first to be taken on strength by the Canadian Forces in 1982 now resides at the Canada Aviation Museum.

above: Although the CF-100 was not quite as fast as its smaller contemporaries, its excellent climbing ability, radar, fire control systems, and twin-engine reliability made it the best all-weather fighter of its time, and ideal for the defence of northern Canada.

MARITIME PATROL AND AIR TRANSPORT

During the Second World War, the RCAF had developed anti-submarine defences in Nova Scotia and Newfoundland. These bases had run down quickly once hostilities ceased. Cold War concerns about Atlantic defences led to the revival of anti-submarine operations, however, and three maritime patrol squadrons were formed between 1950 and 1952. These initially flew Lancasters adapted to sub-hunting (with radar, depth charges, and torpedoes).

The "Lancs," however, were not adaptable to the latest anti-submarine gear and were replaced by Lockheed P2V Neptunes starting in March 1955. Faster than Lancasters and

The CP-107 Argus was a giant among anti-submarine aircraft. It could remain airborne for more than 26 hours, had a range of 7200 kilometres (4,500 miles), and could carry up to 3630 kilograms (8,000 pounds) of weapons such as torpedoes, bombs, depth charges, and mines.

carrying magnetic anomaly detection (MAD) gear in the tails, the Neptunes flew with RCAF squadrons until April 1968. They had a remarkable safety record; not one of the twenty-five Neptunes purchased was lost. They were superseded by the Canadair CP-107 Argus, which entered RCAF service in 1958. At one time Maritime Air Command was simultaneously flying Lancasters, Neptunes, and Arguses. In 1979, the Argus began to give way to the Lockheed CP-140 Aurora.

As one type succeeded another, the number of aircraft deployed decreased steadily. Lancaster squadrons had routinely marshalled twelve aircraft. Argus squadrons regularly operated five aircraft, each with a fifteen-person crew, and Aurora squadrons flew four aircraft, each with an eighteen-person crew. On the other hand, newer machines with more sophisticated equipment were increasingly efficient, as demonstrated in exercises with Canadian, British, Australian, and American forces. In their first twenty-two months with Auroras, No. 405 Squadron reported more submarine contacts than in the twenty-two years of Argus operations.

Operated worldwide by the RCAF, the Canadair North Star gained prominence during the Korean airlift of the early 1950s. Here, one is unloading supplies near Tokyo, Japan, for shipment to Korea.

North Star 17515

THE CANADA Aviation Museum's North Star (No. 17515) was delivered to the Royal Canadian Air Force in March 1948. Prior to the Korean War, it had been engaged in many tasks at home, including support of Red River flood relief operations in 1950. When Operation Hawk began, this aircraft joined the Pacific route. Even a transport airplane could have adventures, and it was no exception. On September 8, 1950, en route from McChord Field, Washington, to Elmendorf Field in Alaska with a full complement of soldiers aboard, one engine failed, then another. Wing Commander C.H. Mussells dumped fuel and made an emergency landing at Sandspit, in the Queen Charlotte Islands, British Columbia. A day later, with passengers and cargo unloaded, Mussells succeeded in taking off on three engines. Final repairs were effected and by September 10, the aircraft was back on duty, embarking with another load for the far side of the ocean.

Until July 1962, 17515 was a participant in No. 426 Squadron's many tasks, from northern resupply to U.N. peacekeeping. It then became a transport trainer. On December 8, 1965, it was one of two North Stars that took part in a ceremonial "stand down" of the type. Two weeks later it became part of Canada's national aeronautical collection and parked at Rockcliffe until it could be housed in the new wing of the Canada Aviation Museum.

In 1946, the RCAF's air transport capacity was limited to what could be carried in a Douglas Dakota transport or towed in a Waco CG-4 cargo glider, neither being adaptable to overseas deployment. Domestic operations consisted largely of shuttling service personnel and freight between bases. The air force acquired transoceanic capability when it acquired Canadair North Stars in 1947.

The Korean War precipitated Canada's first international airlift, when No. 426 Squadron was assigned to Operation Hawk, a four-year action across the Pacific. As RCAF fighter squadrons formed overseas, No. 426 Squadron worked virtually as a scheduled service airline to Europe. The international stage expanded in November 1956 when Canada committed forces to a United Nations Emergency Force in Egypt. This was the beginning of long-term overseas transport work to various locations, including the Congo and Cyprus, where Canadian troops were deployed for years at a time in support of the United Nations.

RCAF air transport was long deployed in domestic mercy missions, ranging from floods in British Columbia and Manitoba (1948) through to the great Ontario and Québec ice storm of 1998. International aid by RCAF airlift commenced in March 1960 when a North Star delivered 3 tonnes of medical aid and eight medical staff to Rabat, Morocco, for delivery to the city of Agadir, which had been devastated by earthquakes. Two months later, No. 426 Squadron's North Stars were delivering relief supplies to earthquake-stricken Chile. International relief operations arising from floods, famines, and natural disasters have become common since the 1960s and were made possible by the development of increasingly larger, long-range transport aircraft.

Nevertheless, the Canadian Forces air transport fleet exists primarily to move combat units into action and to keep them supplied until the mission is completed. The official motto of the Air Transport Group is "Versatile and Ready." High-intensity opera-

facing page: The Museum's Auster A.O.P. 6 (Auster VI) in the markings of a British Army machine flown during the Korean War by Canadian pilot Captain Peter Tees, Royal Canadian Artillery, who was awarded the Distinguished Flying Cross for his actions in directing artillery fire from the air.

tions during the Persian Gulf War of 1991 taxed the fleet, which at that time consisted of twenty-seven Lockheed CC-130 Hercules and five Boeing CC-137s (Boeing 707s), two being tanker aircraft. To wring more flying hours from available machines, commanders delayed inspections and maintenance—unthinkable actions in peacetime. The Boeings were replaced in 1997 by five Airbus CC-150 Polaris (A310) aircraft, two of which have since been converted to a tanker/transport configuration.

ARMY SUPPORT, TEST FLYING, AND OTHER ROLES

The Canadian Army had its own air elements. Early in the Second World War, Britain's Royal Artillery had formed Air Observation Post (AOP) units to scout for the enemy and direct artillery fire, employing light machines such as the Stinson 105 and Taylorcraft Auster.

AOP concepts survived the war. In the Canadian Army, they were most vigorously applied at Rivers, Manitoba, which soon became the focus of tri-service exercises under the general direction of the Canadian Joint Air Training Centre. Army officers took elementary training on types familiar to the RCAF (de Havilland Tiger Moths followed by de Havilland Canada Chipmunks) but then went on to fly the light aircraft that were army specialties: the Auster VI (1947–58; the army bought thirty-four of them for training observers at Rivers, Manitoba) and Cessna L-19 Bird Dog (1955–73). These types were eventually superseded by light helicopters. During the Korean War, Canadian AOP pilots were attached to No. 1903 AOP Flight, a component of the Commonwealth Division.

The development of Canadian military aviation has been accompanied by research in various fields and accomplished by numerous private and government organizations in company with allied nations. Systematic RCAF test flying dated from the establishment of a Test and Development Flight at Rockcliffe in 1931, which originally did little more than check out simple modifications to service aircraft. It evolved into the Test and Development Establishment (1940), which continued its predecessor's work but also dealt with more advanced research, including experiments with aviation medicine and aircraft icing. A Winter Experimental Flight, established at Kapuskasing, Ontario, in 1943, became the Winter Experimental Establishment (WEE) in 1945, with most flying being done from bases at Edmonton, Alberta, and Whitehorse, Yukon. The WEE was a post-war centre for cold-climate testing of American and British as well as Canadian aircraft designs.

above: Crest of the RCAF's Central Experimental & Proving Establishment, mounted on the Museum's Lockheed F-104 Starfighter. Wing Commander Robert A. White challenged the world's altitude record in this machine in 1967.

facing page: The Golden Hawks in one of their famous aerobatic formations during the high point of the RCAF's post-war existence. In the late 1950s and early 1960s, the RCAF was considered a major world air power.

The North offered more than just low temperatures; the solar radiation conditions that produced the aurora borealis also affected communications systems that were by now as important to aviation as the airframes themselves. Thus Churchill, Manitoba, was the centre of a firing range from which research rockets probed the ionosphere. Other work entailed "operational research," the assessment of the effectiveness of equipment either in planned applications or in new roles for which it was adapted. This involved tests of aircraft, ground support vehicles, communications systems, navigational aids, emergency rations, and even thermal socks.

In 1951, the Central Experimental and Proving Establishment (CEPE) was formed, incorporating the former Test and Development Establishment, WEE, and defence components of the National Research Council. Headquarters were in Ottawa, but CEPE operated detachments wherever required, whether in the North or on either seaboard. In 1971, the organization was renamed the Aerospace Engineering Test Establishment (AETE) and moved to Cold Lake, Alberta. There it absorbed other testing units, including No. 448 Test Squadron, Experimental Squadron 10 (a naval unit), and No. 129 Test and Ferry Flight (formerly at Trenton, Ontario). AETE became responsible for the modification and update of current Canadian Forces aircraft while testing new aircraft and proposed technological innovations.

Canadians have trained to be test pilots at centres as diverse as Britain's Empire Test Pilot School at Farnborough and the USAF Test Pilots School at Edwards Air Force Base, California, the latter having trained Canadian astronaut Colonel Chris Hadfield. Instruction at these centres involves flying numerous and varied types of aircraft, old and new, with an emphasis on how to adapt quickly to unfamiliar instruments, cockpit layouts, and flight characteristics.

Throughout the Cold War years, the RCAF/Canadian Armed Forces maintained a public profile through aerobatic performances at air shows and ceremonial events. Between 1929 and 1934, a flight of Armstrong Whitworth Siskin fighters performed at events ranging from the Cleveland Air Races to celebrations of Toronto's centennial in 1934. Following the war, the aerobatic team was succeeded by a series of air demonstration teams, some of which were short-lived.

Several squadrons and schools formed aerobatic teams of varying duration, flying whatever current equipment was at hand, including North American P-51 Mustangs, *facing page:* Wing Commander Robert A. White entering his Lockheed F-104 Starfighter in 1967 (see also the photo on page 132). His mission with the RCAF's Central Experimental & Proving Establishment entailed twelve flights above 30 000 metres (98,500 feet).

Canadair T-33 Silver Stars, and McDonnell F2H-3 Banshees. Their public appearances, however, were secondary to their training or operational functions. In 1959, the RCAF formed the Golden Hawks (flying Canadair Sabres) to celebrate the fiftieth anniversary of powered flight in Canada. Its sole function was aerobatic performances across the nation. It was expected the team would be disbanded at the close of the year, but its popularity (and obvious public-relations benefits) ensured its survival until 1964, by which time the team had appeared at 317 air shows.

Canada's centennial celebrations in 1967 included a new aerobatic team, the Golden Centennaires (flying Canadair CT-114 Tutors), which appeared at the opening and closing ceremonies of Expo 67 and on ninety-eight other occasions. The group disbanded at the close of 1967, and three years passed before another Tutor aerobatic team appeared. Formed at Moose Jaw, Saskatchewan, initially with three aircraft, it seemed to be yet another unit-based formation with a short anticipated life. In 1971, it was expanded to seven aircraft and given a name—the Snowbirds—and has been so consistently popular with the public that, as of 2005, the team continues to perform.

After the Second World War, the world experienced decades of political stagnation during the Cold War. An equilibrium maintained under the ominous but stabilizing shadow of nuclear confrontation between Western and Eastern power blocs averted global warfare but did not prevent the outbreak of regional conflicts. The policy of mutual standoff between the superpowers permitted, indeed encouraged, the perpetuation of localized conflicts such as the Korean War, the Vietnam War, successive Arab–Israeli confrontations, and the civil wars of the 1980s in Afghanistan, Lebanon, and Nicaragua, as well as conflicts across Africa.

The symbolic sign that the Cold War had run its course was the breaching of the Berlin Wall in 1989. With communism discredited and the world unleashed from the inhibitions of superpower polarity, political analysts may have been seduced into believing that a new world order was about to be established. Yet the euphoria has proved premature. The removal of the Cold War's East–West equilibrium left a political vacuum, opening the way to a resurgence of antagonisms, mostly based on ethnic nationalism, that had been hidden beneath the artificial stability of the Cold War balance.

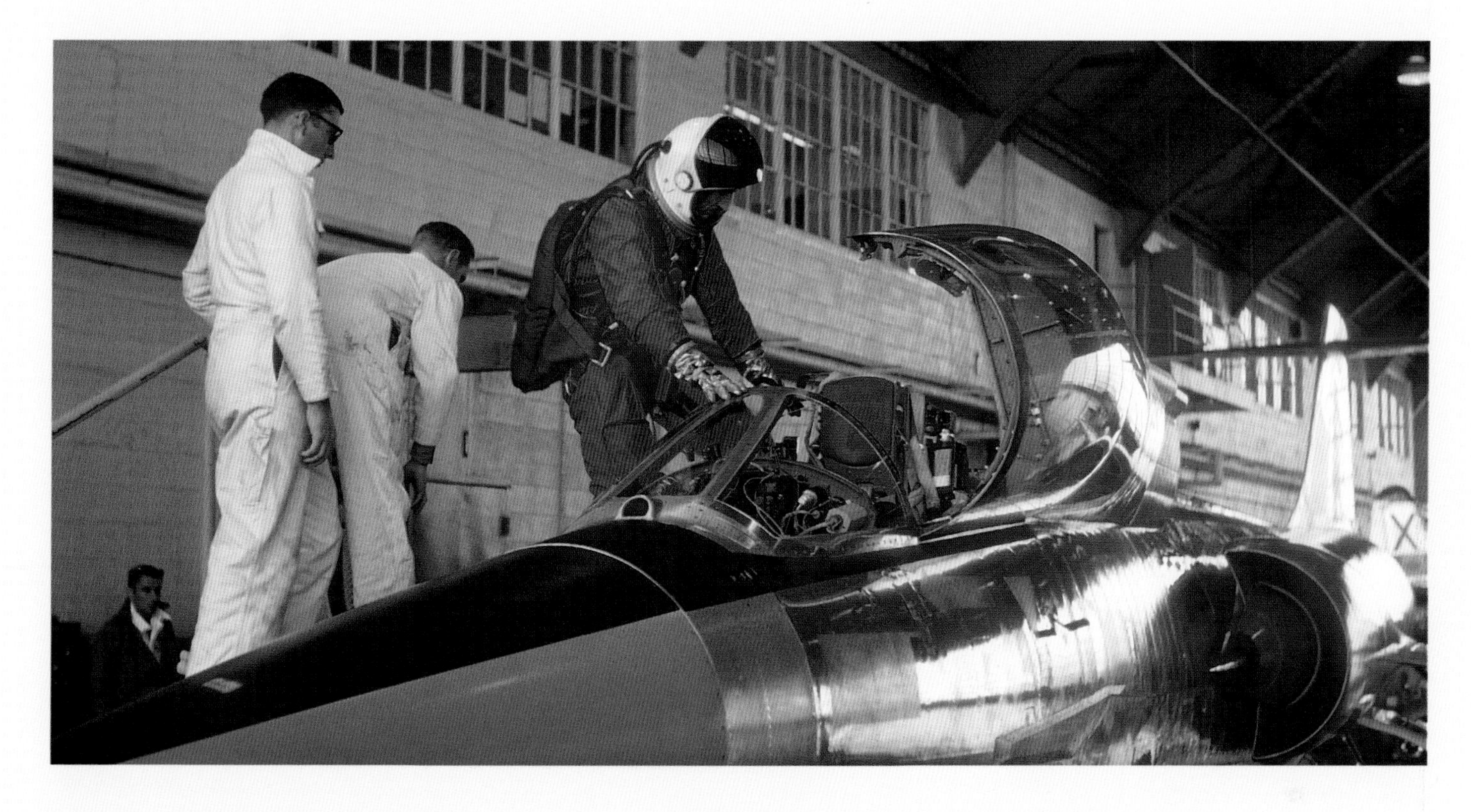

A Canadian Altitude Record

TEST FLYING has always mixed success and failure, sometimes simultaneously. The Canada Aviation Museum's F-104A, 104700, was used by Wing Commander Robert A. White on December 14, 1967, to establish a Canadian altitude record of 30 510 metres (100,097 feet), attained by alternately diving (to build up speed) then climbing. It had, however, been an attempt to beat a Soviet record of 36 740 metres (120,536 feet). He later described the experience as being "like a fly riding on an artillery shell. I could control the altitude but not the trajectory." At 24 200 metres (79,395 feet) the afterburner failed, and at 27 100 metres (88,909 feet) he shut down the engine to prevent it overheating. Although the prime goal was not achieved, White's mission entailed twelve flights above 30 000 metres (98,500 feet), a unique achievement at the time.

LIMITED
CANADIAN AIRWAYS
LIMITED
Good for ONE FLIGHT
AIR
CANADA
TRANS-CANADA
AIR LINES
TRANS-CANADA Air Lines
T.C.A. FOUNDERS
To commemorate the founding in 1937 of Canada's transcontinental air line, the following persons active in its beginning have signed
AIR CANADA
CANADIAN PACIFIC
AIR CANADA

the DEVELOPMENT *of air* TRAVEL

8

facing page: A selection from the Museum's collection of artifacts and memorabilia that date from the early years of airline development to the present day.

AIR TRAVEL AND transport are now so common that we accept many of their benefits unthinkingly. Apart from business or pleasure, these benefits may be as mundane as the speedy movement of New Zealand kiwi fruit to Canadian breakfast tables, and as vital as the rapid movement of donated organs to waiting recipients. It was not always so.

In March 1913, Igor Sikorsky pointed the way to modern airliners with his giant, twin-engine Russkii Vityaz (Russian Knight), nicknamed *Le Grand*. It was large and comfortable, with an enclosed cabin and even a toilet for passengers. It was also under-powered, and could climb no higher than 100 metres (330 feet). Sikorsky enlarged it, fitting it with four motors. The resulting *Ilya Muromets* design (named after a mythical Russian hero) set records for size and performance, flying from St. Petersburg to Kiev and back. War interrupted commercial development everywhere, and Russian aeronautics in particular was set back by revolution and then civil war. Sikorsky fled to the United States in 1918, where he resumed his aviation career—but the promise of the *Ilya Muromets* took decades to realize.

In the immediate aftermath of the First World War, nations scrambled to find commercial applications for aircraft, including scheduled airmail, freight, and passenger services. The first "air liners" were converted bombers with no pretence of comfort. Finding a niche for air travel was difficult, as the early transports lacked lifting capacity, speed,

above: The interior of a Vickers Vimy Commercial airliner, *c.* 1919, shows that early flights on Imperial Airway's London to Paris route offered no pretence of passenger comfort.

facing page, top: Igor Sikorsky's giant twin-engine *Le Grand* (1913) pointed the way to modern airliners, but its promise took decades to realize.

facing page, bottom: The first airliners were converted bombers such as this Vickers Vimy Commercial. Developed from the Vimy bomber of the First World War, it had a range of 725 kilometres (450 miles) and carried ten passengers.

and range, and could not compete with the railway systems found in Europe and North America. European airlines evolved only with large government subsidies.

Airliners evolved to meet the special needs of different clienteles, including international prestige and a wish to link imperial capitals with distant protectorates. Technology advanced slowly. Biplanes, with accompanying struts and bracing wires, persisted throughout the 1920s. Even beautifully streamlined Curtiss Condors used the biplane layout. Some European airliners seemed to defy modernity. The huge Armstrong Whitworth Argosy and Handley Page Hannibal were elegant inside but aerodynamically inefficient.

Early airline designers faced many problems. Single-engine machines had limited lift capability; two engines enabled larger loads but brought no appreciable performance benefits. The result was the appearance of numerous tri-motor designs in the United States and Europe.

The modern airliner began to take shape in the early 1930s as several technological advances appeared and converged. Powerful and reliable engines were part of the equation, but so were the perfection of cantilever wings (with no high-drag external bracing, wires, or struts), monoplane layouts, retractable undercarriages (which worked best on low-wing monoplanes), advanced streamlining (refined in the course of numerous aerial speed contests), and stressed-skin metal construction using lightweight aluminum. Controllable-pitch propellers offered improved performance and attained commercial practicality. Sound-proof cabins were developed about 1931 and greatly improved passenger comfort.

An important milestone in air transport history occurred in 1933, when the Boeing 247 was introduced. It was a ten-passenger aircraft of revolutionary concept, and it may be considered the world's first modern airliner, having almost all of the basic design features that became standard on later transport aircraft. This all-metal, streamlined

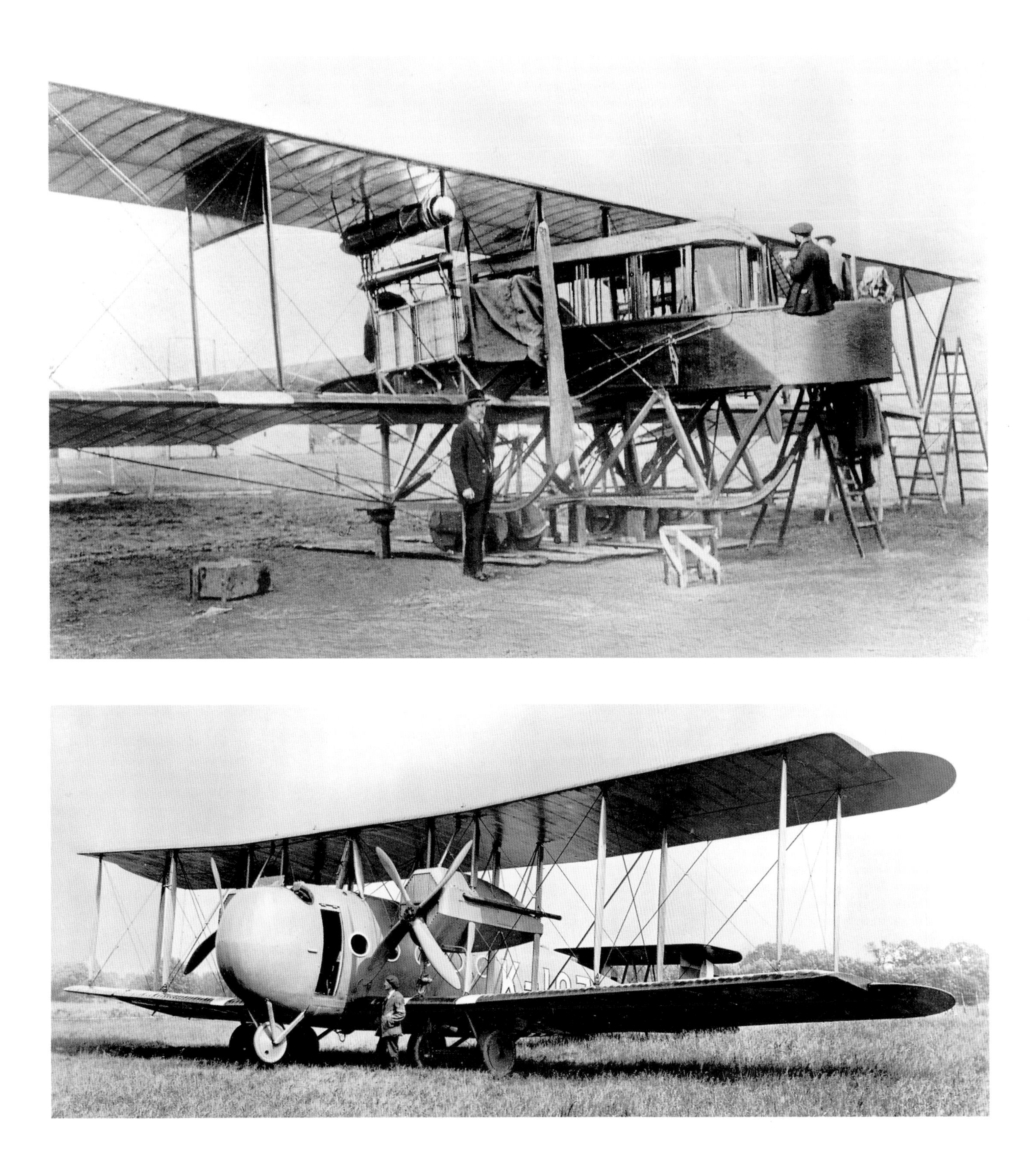

Wallace Rupert Turnbull
and the Variable-Pitch Propeller

PROPELLER PITCH is the angle at which blades slice the air. Efficiency varies according to the aircraft's altitude and attitude (whether climbing or flying level), and the demands of takeoff differ from those of cruising. In 1916, Wallace Rupert Turnbull (1870–1954) of Rothesay, New Brunswick, began investigating how propeller pitch could be altered by a pilot while in flight. His later designs used a small electric motor to alter the blade angle. At Camp Borden, Ontario, 1923 ground tests with a mechanically activated propeller fitted to an Avro 504 were followed in 1927 by air tests of the electrically activated design shown above. Within two years Turnbull had sold his patents to the Curtiss Wright Corporation and to the Bristol Aeroplane Company.

monoplane incorporated a semi-monocoque fuselage, neatly cowled twin engines, and an enclosed cockpit for the pilots. The undercarriage was retractable and the improved Model 247D had controllable-pitch propellers. With a maximum speed of 290 kilometres per hour (180 miles per hour), it set the standard for subsequent airliner development and was the aircraft that inspired the design of the Douglas DC series of commercial transports.

The new technologies included enormous improvements in radio communications for both navigation and bad-weather flying. Unhappily, increased flying in clouds meant greater hazards from icing, which led in turn to the development of heated carburetor air intakes and rubber de-icing boots on the leading edges of wing and tail surfaces.

Many of these technical innovations pioneered by the Boeing 247 and other aircraft were incorporated into a rapid succession of attractive American airliner designs: the Douglas DC-2 (1934), the Lockheed 10 Electra (1934), the Lockheed 12 Electra Junior (1936), and the Douglas DC-3 (1936). Their European competitors trailed badly and often found favour only with national airlines compelled to buy the national product. As of 1937, only the angular, tri-motored Junkers JU 52/3M, which had appeared in 1932 and retained the fixed undercarriage, competed with the American models in airliner sales.

The most important of these American airplanes was the DC-3, which was developed as an enlarged and improved version of the DC-2, with berths for night flights. The resulting wider fuselage and longer wing span made the DC-3 even more economical as an airliner and adaptable to cargo. Including the military C-47 version, known as the Dakota in the Royal Canadian Air Force (RCAF) and the Royal Air Force (RAF), some 13,500 were manufactured (2,500 under licence in the Soviet Union and Japan) and approximately 3,500 were still flying in 1990. The DC-3 cruised at 270 kilometres per hour (170 miles per hour) with twenty-one passengers or 3 330 kilograms (7,300 pounds) of freight. Their success and longevity lay in durability, adaptability, economy of operation, and sheer numbers manufactured.

above: Model of the all-metal Ford 4-AT Trimotor, introduced at the end of the barnstorming era. It was America's first successful airliner.

facing page: Detail of Turnbull's first electrically operated variable-pitch propeller, mounted on the Museum's Avro 504K.

Canadian air transport companies, long tied to frontier operations, struggled during the 1920s with limited passenger traffic and irregular freighting contracts. "Scheduled services" aimed merely at flying on a particular day, seldom at a predictable hour. The Curtiss HS-2L flying boats and de Havilland DH.9A machines that Laurentide Air Service used in Québec between Angliers, Lac Fortune, and Rouyn from May 1924 were adequate to meet mining needs but were not airliners. In their quest for revenues, air carriers transported airmail, charging whatever fees they could. Firms like Patricia Air Service even printed and sold their own airmail "stamps" for letters and parcels.

In 1927, the Post Office awarded airmail contracts to a few companies for specific routes and began printing special airmail stamps. Competition between carriers drove down the cost of air postage. Some agreements were for weekly deliveries. On the other hand, winter deliveries to communities along the north shore of the St. Lawrence River, for instance, were seasonal and sporadic, depending on weather and the amount of mail. The airmail contracts gave air carriers some stability by enabling them to predict their revenues and forced them to strive for more precise scheduling.

Passenger and airmail services grew dramatically, and by 1930 numerous airmail services were in operation six days a week in both Central and Western Canada. Even so, air schedules were set to reflect other factors, including Canadian railway timetables and the departure of mail flights from American airmail hubs.

Prior to the Second World War, approximately 90 per cent of scheduled American passenger flights were made with Douglas DC-3s, and large numbers of its military version were later converted into airliners, such as this one in Canadian Pacific Airlines service.

Unhappily, airmail contracts proved less predictable than had been hoped. In 1931, as the Depression took hold, the Post Office cancelled many contracts and curtailed the scope of others. By the spring of 1932, airmail routes had shrunk to a few northern runs and to isolated communities such as Pelee Island in Ontario—the routes that could not adequately be replaced by land-based delivery.

A CANADIAN TRANSCONTINENTAL AIRLINE

The early American airlines had struggled during the 1920s, relying not on passenger traffic but on mail contracts for survival. The new developments in airliner designs, with their improved safety, comfort, speed, and reliability, made possible the evolution of airline companies as we now know them in the United States and Canada. Plans for a Canadian national air transport system were put on hold by the Great Depression, yet simultaneously the foundations for such a network were laid. To employ thousands of young men out of work, the government of R.B. Bennett authorized a series of relief projects (until they were phased out in 1936), most of them supervised by military officers. They included construction of emergency airfields across the nation as part of a proposed Trans-Canada Airway. Although few of the airfields ever handled an airliner, many became bases or relief fields for Second World War flying schools.

But who would operate along a new transcontinental airway? James Richardson hoped that his Canadian Airways would become the national carrier. In 1936 the Department of Transport was formed by merging the Department of Marine with the Department of Railways and Canals and removing from the Department of National Defence all responsibilities for civil aviation. The new minister of transport, Clarence D. Howe (1886–1960), suggested an airline jointly owned by Canadian Airways, Canadian Pacific Railway, and Canadian National Railway. Unable to get three-way agreement, Howe went ahead and, on April 10, 1937, Trans-Canada Air Lines (TCA) was formed as a subsidiary of the (government-owned) Canadian National Railway.

Trans-Canada Air Lines began modestly; its first route was a Vancouver–Seattle service, operated with a Lockheed 10A Electra aircraft and crew, both formerly of Canadian Airways. Neither the Lockheed 10A nor the later Lockheed 12 Electra Junior were suitable for long routes. TCA acquired Lockheed 14 Super Electras and gradually expanded its routes until by 1941 it was operating from Vancouver to Halifax.

Lockheed had acquired their expertise in low-wing monoplane transports and retractable undercarriages in their Model 9 Orion—a wooden, single-engine machine. In 1932, following reorganization of the company, they abandoned single-engine concepts, gambling on two engines and metal construction. The resulting Lockheed 10 Electra, first flown in February 1934, was the first in a series of successful designs. In all, 149 of them were built.

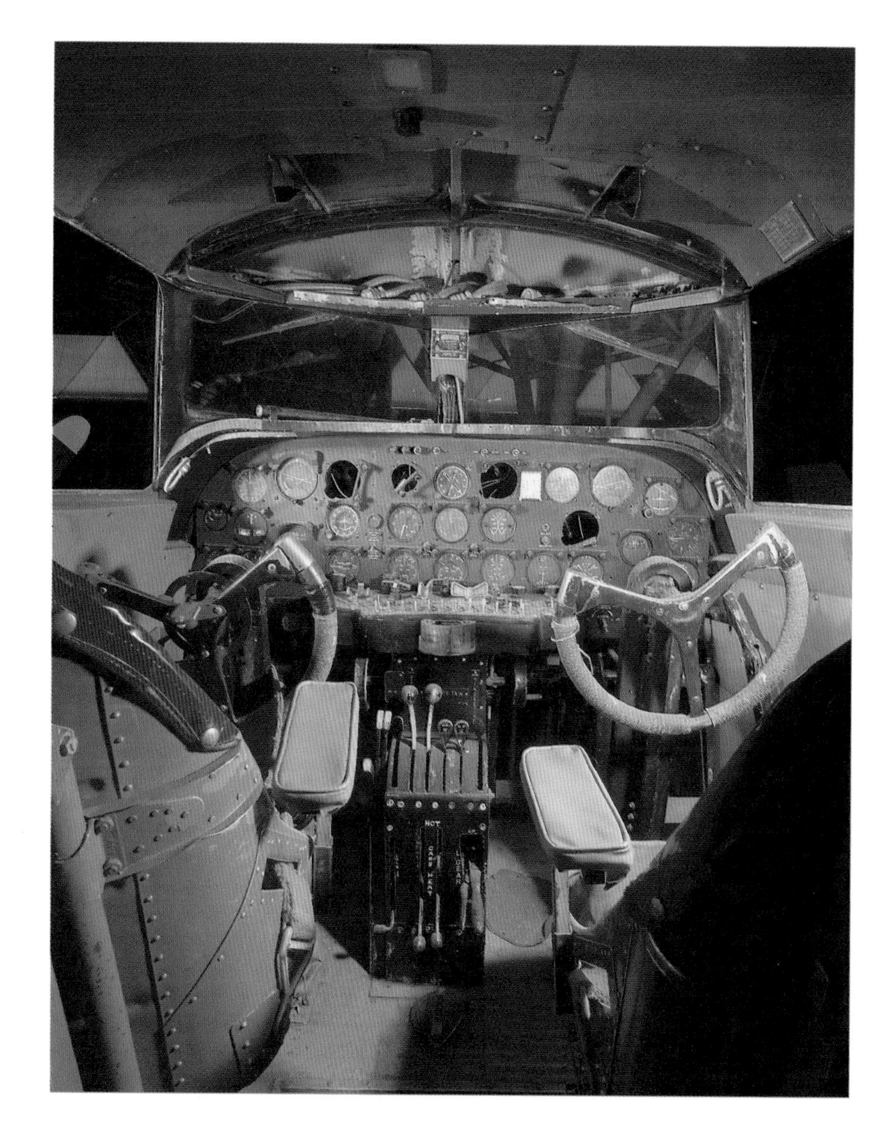

Cockpit of the Museum's Boeing 247, ancestor to the modern airliner. Only 75 were built, as it was quickly surpassed by the larger Douglas DC-2. Some people at Boeing thought the 247 should have been designed larger, but company pilots wanted a smaller aircraft that would be more manoeuvrable and easier to land.

The creation of Trans-Canada Air Lines led directly to the formation of its rival. Canadian Pacific Railway had held Canadian Airways stock since 1930, but in 1939 it studied the possibility of having its own air carrier. The next year it bought out ten struggling local carriers, and in 1942 they were amalgamated as Canadian Pacific Airlines (CPA). One of the constituent companies, Yukon Southern Air Transport, furnished its most renowned president, George William Grant McConachie.

Within Canada, both companies marked time during the Second World War (though transoceanic flying was another matter). Immediately after hostilities, however, they set out to modernize their fleets. Old aircraft were either sold or shuffled to secondary routes. Both companies then bought Douglas DC-3s from large war-surplus stocks. Next they acquired four-engine Canadair North Stars, TCA for both the Canadian and trans-Atlantic market, CPA for trans-Pacific operations.

James Armstrong Richardson (1885–1939): A Father of Canadian Aviation

A WEALTHY GRAIN merchant in Winnipeg, James Richardson founded Western Canada Airways in 1926. Four years later, following company mergers, he controlled Canadian Airways, a decentralized firm operating in all parts of Canada, which he hoped to integrate into a single transcontinental mail and passenger network. Even after the government cancelled most of its airmail contracts in 1931, Richardson clung to the hope that he would eventually preside over a national airline.

When the Bennett government was defeated in 1935, Richardson expected transcontinental air route plans to be revived, and he knew that Canadian Airways would be the logical company to execute those plans. Negotiations proved complex, involving railway and air transport companies, the Post Office, and the new Department of Transport. Richardson was intractable about sharing power with any other party. The opportunity to establish a national airline eluded him in his lifetime, but in 1975 he was inducted posthumously into Canada's Aviation Hall of Fame, which declared that "in the annals of this nation's flying history, no businessman gave more of himself for less reward to the everlasting benefit of Canadian aviation."

Aircraft challenged both distances and natural barriers. The ultimate obstacles were the Atlantic and Pacific Oceans. At first, only adventurous pioneers attacked them. For some, like Australian Charles Kingsford-Smith and American Charles Lindbergh, there was success and fame. For others, transoceanic flights ended in privation and death.

Initially, hopes for transoceanic air travel lay with airships. Three weeks after the British flyers John Alcock and Arthur Whitten Brown had made the first, hair-raising, non-stop transatlantic flight (in a converted Vickers Vimy bomber) from Newfoundland to Ireland on June 14–15, 1919, the second non-stop transatlantic flight was made—in considerably more comfort, if with less speed—by an airship, Britain's R-34. By completing the journey it also made the first crossing from east to west, and on returning, became the first flying machine to cross in both directions.

However, accidents plagued military airships throughout the 1920s and undermined confidence in commercial airships. Nevertheless, Great Britain planned to use them as links to the empire. Mooring towers were built at St. Hubert near Montréal, Ismailia in Egypt, and Karachi in India (now in Pakistan). The R-100 flew to Canada and back in July and August 1930. At 212 metres (695 feet) long, the R-100 was more than twice the length of a football field. Among those associated with its design were N.S. Norway (better known as author Nevil Shute) and Barnes Wallis (designer of the Vickers Wellington

In *Time Warp*, a painting commissioned by Air Canada for its fiftieth anniversary, Robert Bradford portrays a Lockheed 10A, one of the first modern airliners to be introduced into Canada by Trans-Canada Air Lines in 1937. This particular aircraft was restored to airworthiness by the airline in the 1980s and flown for charity purposes. In the background is a contemporary Boeing 767 wide-bodied jet.

A McGraw-Hill Publication NOVEMBER, 1933 Price 35c. per copy

AVIATION

The Oldest American Aeronautical Magazine

PRATT AND WHITNEY ENGINES *Exclusively*

POWER THE PLANES THAT FLY 46,400 MILES DAILY ON REGULAR SCHEDULE OVER UNITED AIR LINES ROUTES

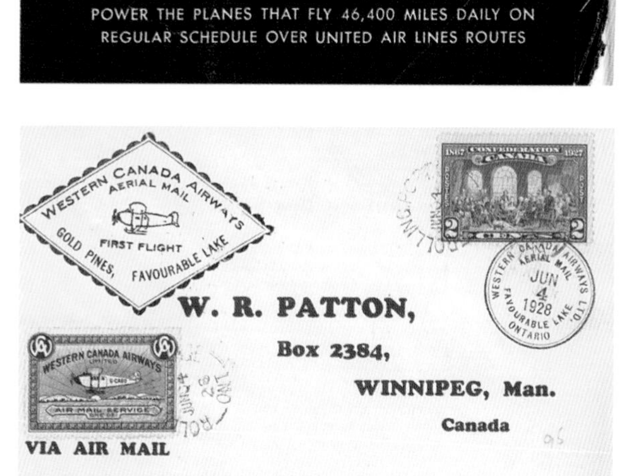

bomber and of the bouncing bombs that breached German dams in 1943). The highlight of the airship's career was its visit to Canada in the summer of 1930, which was a media event in itself. An estimated one million people made their way to St. Hubert to view the airship, although only about three thousand ascended the mooring mast to tour it. The ship's officers dined with the Governor General, delivered speeches to Toronto's Empire Club, and generally promoted airship travel as something deserving support from the Dominions as well as Britain. On August 10–11, the R-100 flew over Ottawa, Toronto, and Niagara Falls; Howard Hughes was reported to have offered $100,000 to have it visit New York.

But dark clouds threatened. On October 5, 1930, the R-100's sister ship, the R-101 (a completely different design), crashed in France en route to India, killing forty-eight of the fifty-four persons aboard. Among the dead were many of the most enthusiastic proponents of airship travel. The disaster effectively ended the British airship program.

Germany had a successful commercial airship program and, like Britain, was handicapped by reliance on flammable hydrogen rather than inert helium. From 1928 to 1937, their LZ127 *Graf Zeppelin* made 590 flights and carried 13,100 passengers without injury, principally between Germany and Brazil. The *Hindenburg* (LZ129), designed for airship service to the United States, made 11 trips in 1936, but burned while docking at Lakehurst, New Jersey, on May 6, 1937, killing 35 of the 97 persons aboard. The cause of the accident has never been firmly established, but it was remarkable for having been caught by news cameras, creating one of the twentieth century's iconic images. Yet even if the Hindenburg had not crashed, it is doubtful that such travel would have lasted beyond 1939; first war and then great strides in airliner development would have ended the era of the stately, luxurious airship that was, in effect, the Concorde of its day.

The next best hope for scheduled transoceanic air travel was the flying boat. British designers long clung to biplane patterns; they almost literally designed "boats that could fly." American designers followed another route—transports that could operate from water—exemplified in the Sikorsky S-38, S-40, and S-42. The British began to catch up only when they belatedly adopted the American approach. Even so, the Empire flying boats operated by Imperial Airways were fragile compared to the rugged Boeing 314s of Pan American Airways. Commencing in 1939, Foynes in Ireland and Botwood in

facing page, top left and bottom right: In 1927, the Canadian Post Office awarded airmail contracts to a few companies for specific routes and began printing special airmail stamps. Illustrated are first-day covers collected and later donated by the late Kenneth M. Molson, the Museum's first curator.

facing page, top right: A Boeing 247 is featured on this cover of *Aviation* from November 1933. An all-metal, twin-engine, low-wing monoplane with a retractable undercarriage, it ushered in the era of the modern airliner.

facing page, bottom left: Biplane airliners, with their external struts and bracing wires, persisted well into the 1930s. This advertisement in *Aviation*, July 1934, shows an American Airlines Curtiss Condor, a "luxurious new convertible sleeper plane" of the day.

George William Grant McConachie (1909–1965): Bush Pilot with a Briefcase

GRANT MCCONACHIE expanded commercial air transport from Canada's northern frontiers to the international stage. Growing up in Edmonton, he was a frequent visitor to the local airfield, where he did odd jobs for famous pilots like Punch Dickins and Wop May for a chance to fly with them. By age twenty-two, he had his commercial pilot's licence and was on his way to China to find work when he visited an uncle in Vancouver who, in an effort to entice his nephew to remain in Canada, provided the financing for a second-hand Fokker Universal to start his own airline.

Although forced to close his first company after a crash hospitalized him, in 1933 he became a partner in a new firm called United Air Transport, later renamed Yukon Southern Air Transport, a future component of Canadian Pacific Airlines.

By 1939 he had pioneered the first scheduled, dependable airmail and passenger service between Edmonton, Alberta, and Whitehorse, Yukon, despite the near-insurmountable obstacles of weather, inhospitable terrain, and mechanical difficulties. Since his routes were found to be the best way of reaching farther north, the completion of the Alaska Highway and of airfields for the North West Staging Route that delivered Allied aircraft to the Soviet Union during the Second World War were achieved much earlier than would otherwise have been possible.

During the war, he was named general manager of western lines in the newly formed Canadian Pacific Airlines (CPA). Wartime work in the North and his effective merging of these diverse companies earned him the McKee Trophy for 1945. Appointed president of CPA in 1947, he established services to Asia and South America as well as a polar route between Canada and Europe. Through imaginative expansion into the jet age, he astutely kept CPA profitable to the day he died.

Newfoundland briefly became focal points for international air travel. By the year's end, a new world war had disrupted commercial development. When peace returned in 1945, the long-range land plane was well on the way to displacing the flying boat.

In 1940–41, the North Atlantic Ferry Organization was created by the Royal Air Force and the Canadian Pacific Railway. Its task was to deliver, by air, the hundreds, then thousands of aircraft flowing from North American factories to Allied bases overseas. The Canadian nexus for Ferry Command (as the organization became) was Dorval, near Montréal, and new bases were developed in Newfoundland, Greenland, Iceland, the Azores, and West Africa as stepping stones to Britain. With the bases came sophisticated radar and radio navigational aides.

The routes traversed by ferried aircraft were joined in 1943 by transport aircraft operating in accordance with loose schedules. British Overseas Airways Corporation (BOAC) led the way in 1941 with Consolidated Liberators bringing aircrew back to Canada. TCA looked enviously at the North Atlantic route, but had to wait until converted Avro Lancaster bombers made it feasible. Operating under another name (Canadian Government

above: Over the Coast, by Frank Oord. In 1919, the trans-Atlantic crossing of the huge R-34 dirigible tested the potential of the rigid airship as a passenger carrier. At the time, the flight seemed to be a natural progression in the onward march of aviation.

facing page: A young Grant McConachie stands in the open cockpit of a Fokker Universal belonging to Western Canada Airways. At the age of twenty-two, he started his own airline with this type of aircraft.

above: Converted Lancaster bombers completed hundreds of trips for the Canadian Government Trans-Atlantic Air Service during the Second World War. Later, Trans-Canada Air Lines used them to provide Canada's first trans-Atlantic passenger service open to the travelling public, *c.* 1946.

facing page, top: Canadian Pacific Airlines inaugurated its passenger service to the Far East with the Canadair North Star in 1950.

facing page, bottom: In the early 1950s, Trans-Canada Air Lines replaced its Canadair North Stars on the Atlantic route with Lockheed Super Constellations to take advantage of the post-war flood of immigrants and the increasing business of holiday travel.

Trans-Atlantic Air Service), they made their first crossing from Dorval to Prestwick, Scotland, in July 1943. The most important loads were mail sacks rather than passengers, but the service was inadequate at a time when Canada had over 500,000 service personnel overseas. To speed things along, the RCAF formed No. 168 (Heavy Transport) Squadron to carry mail overseas. It commenced operations from Rockcliffe in December 1943 using Boeing B-17 Fortresses, which in August 1944 were joined by Liberators. In twenty-eight months, they made 636 round trips across the Atlantic and moved nearly 1 000 tonnes of mail.

No. 168 Squadron was a temporary wartime service, but TCA intended to stay on the Atlantic routes. More Lancaster transports joined the fleet, and in August 1944 they began serving cold meals to their government passengers. In the summer of 1945 they offered tickets for sale to the general public. Nevertheless, the aircraft were still just converted bombers carrying uneconomical loads with minimal comforts.

POST-WAR DEVELOPMENTS IN AIRCRAFT AND AIRLINES

Wartime production had left huge surpluses of medium-sized aircraft. For decades, Douglas DC-3s and Curtiss C-46 Commandos serviced domestic air routes and carried air freight throughout the world. These durable, ubiquitous aircraft made it difficult for manufacturers to introduce new medium-haul designs into the market.

434
CF-CPP
Canadian Pacific Airlines
TRANS-CANADA AIR LINES
403

Gordon R. McGregor (1901–1971)

As a boy, Gordon McGregor witnessed the birth of aviation in Québec at the Montreal Air Meet of 1910. Later in life, he joined a flying club in Kingston, Ontario, winning competitions for the Webster Memorial Trophy—awarded annually to the best amateur pilot in Canada—in 1935, 1936, and 1938.

With war on the horizon, McGregor joined the air force and in 1939 proceeded overseas with No. 1 Fighter Squadron, Royal Canadian Air Force (RCAF). At the age of thirty-nine, he was the oldest Canadian fighter pilot to serve in the Battle of Britain. Noticed for his leadership and management skills, McGregor returned to Canada and was given various commands on the west coast to assist in the development of fighter operations following the Japanese attacks in the Aleutians off Alaska. Recalled to Britain in 1944 to prepare for the invasion of Normandy, a wing led by McGregor was the first to base itself permanently on the continent after D-Day. For his exceptional services in the European theatre, McGregor was named an Officer of the Order of the British Empire (OBE).

Following the war, McGregor was hired by Trans-Canada Air Lines (TCA) and in 1948 was named its president. In the early 1950s, McGregor knew the future lay with turbine engines, and in 1955 it was a particularly bold step to opt for the revolutionary Vickers Viscount turboprop, making TCA the first North American airline to operate turbine-powered aircraft. With the arrival of the Douglas DC-8 turbojet in the early 1960s, TCA could claim to be the first airline in the world whose fleet consisted entirely of turbine-powered aircraft.

When transport minister C.D. Howe appointed McGregor president of the company, he was reported to have said, "You stay out of the taxpayer's pockets and I'll stay out of your hair," to which McGregor replied, "I'll have you out of my hair as soon as possible." In the twenty years until his retirement in 1968 the company lost money only twice.

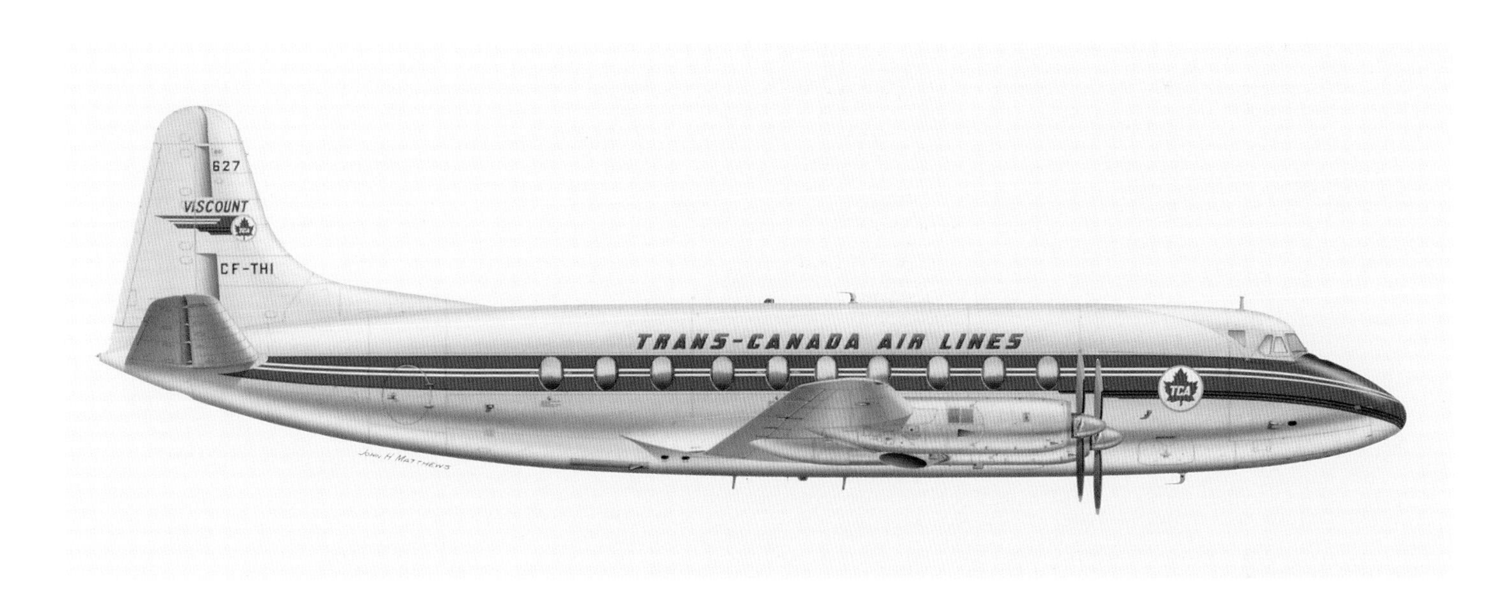

above: Trans-Canada Air Lines flew fifty Vickers Viscounts on domestic routes between 1955 and 1974, the first turbine-powered passenger aircraft to be operated in North America. Retired by Air Canada in 1969, the Museum's example was restored to its original TCA markings before presentation to the Museum.

facing page: Wartime portrait of Gordon R. McGregor wearing the rank insignia of an RCAF Group Captain.

It was a different matter on long-haul routes. Aircraft that were advanced in development in 1945 became common airport visitors by 1950. The Douglas DC-4 and Lockheed Constellation made non-stop transcontinental and transoceanic air travel both comfortable and economical. Newer aircraft followed quickly, with the DC-4 giving way to the DC-6 and DC-7 and the Constellation to the Super Constellation—and the Boeing Stratocruiser competing with both. Yet these machines in their turn gave way relatively quickly to jet transports such as the de Havilland Comet, the Boeing 707, and the Douglas DC-8.

As war-surplus airliners were retired from short-haul domestic routes, they were replaced by medium-sized turboprop machines (such as the Vickers Viscount and Vanguard) or the jet aircraft with which we are still familiar (such as the Douglas DC-9 and Boeing 737). The DC-9 proved to be the longest-serving aircraft type in the Air Canada fleet, flying from 1966 to 2002. Robert Giguère, who piloted the type for ten years before becoming an Air Canada executive, said: "The DC-9 changed the nature of air transportation in Canada and represented for most Canadians their first experience of jet travel. It connected Canadians from coast to coast, between most Canadian cities large and small, with faster, quieter, and more reliable air transportation than ever before."

The 707s and DC-8s that so easily displaced the previous generation of long-haul conventional airliners were in turn challenged and then supplanted by "jumbo jets" like the Boeing 747, the Lockheed L-1011, the Douglas DC-10, and the various models of Airbus. At the beginning of the twenty-first century, the pattern is set to repeat itself as the next generation of "superjumbos," led by the Airbus A380, comes into service.

The most obvious changes in design through these generations of post-war airliners have been the introduction of efficient turbofan engines, swept wings, and increased size, speed, and range (making a stop at Gander in Newfoundland or Goose Bay in Labrador is no longer mandatory for transatlantic flights). Inside the sleek fuselages, the introduction

By the 1950s, increasing numbers of passengers per aircraft led to the hiring of more flight attendants, such as this 1954 graduating class of Trans-Canada Air Lines stewardesses.

of avionics permitted reduction of the flight crew; a Super Constellation had needed at least five aircrew, whereas one of the modern Airbus types normally requires no more than two. Increasing numbers of passengers per airplane led to more flight attendants (formerly stewards and stewardesses), one of whose tasks was to serve meals in flight. Flight Data Recorders and Cockpit Voice Recorders, widely used since the mid-1960s, have increased airline safety through scientific accident investigations. Since the 1980s, the greatest changes to airliners have been the introduction of digital cockpit displays, "fly-by-wire" technology, and use of composite materials.

The Canadian air transport industry went through many changes following the Second World War. By 1947 Trans-Canada Air Lines was most visible on transcontinental and transatlantic routes. Its new rival, Canadian Pacific Airlines, was initially occupied with short-range domestic services. In 1949 CPA commenced trans-Pacific operations, losing money until the outbreak of the Korean War, when flying to the Orient and South Pacific became more intense. Major corporate restructuring in 1986 led to CPA being absorbed by a new entity, Canadian Airlines International. In 1965, TCA had been renamed Air Canada, and since 1962 there had been frequent suggestions that Canada's two international flag carriers should merge. Airline deregulation in 1987 increased competition among all carriers, many of which had to reorganize or merge to survive. Finally, Air Canada took over Canadian Airlines International in 2000.

A handful of local carriers from the 1940s grew into substantial regional airlines. The story of Canadian air transport carriers is complex and littered with the names of now-vanished companies such as Air Alberta, Norcanair, Transair, Nordair, and City Express, as well as charter specialists like Wardair. Nevertheless, assorted "firsts" and other milestones stand out. For example, the far North, with its short, gravel runways, was long the

In the unpressurized cabin of a TCA Lockheed Lodestar, an oxygen mask was often a necessity when crossing over the Canadian Rockies in the early days of transcontinental service. Of note is the military look of the flight attendant's uniform, *c.* 1943.

above: Air Canada's longest-serving DC-9 was retired to the Canada Aviation Museum in 2002. Here, former Air Canada captains pose with the plane in a group portrait with delivery pilot Robert Giguère (standing third from left), whose father also flew DC-9s with the company.

facing page: Max Ward's first company, Polaris Charter Company Ltd., based in Yellowknife, Northwest Territories, began with this de Havilland Fox Moth. Ward is pictured with the aircraft *c.* 1947, in which he hauled prospectors and supplies into the mining and exploration camps of the North.

domain of propeller-driven aircraft, but in March 1969 Nordair instituted jet services to the High Arctic using Boeing 737 jets. In 1973, Transair hired Rosella Bjornson as the first female airline pilot in Canada and the first woman in North America to be qualified as a first officer on jet transports. In 1984, Eastern Provincial Airways and Torontair (a short-lived local company) were the first airlines to institute non-smoking policies throughout their fleets. The first scheduled jet services between the eastern and western Arctic came to pass in 1988, and Bradley Air Services/First Air thus became Canada's least-known transcontinental airline.

Charter flights (as distinct from scheduled airline services) evolved after the Second World War. From 1962, they became increasingly popular with tourist agencies. Whereas established carriers used charter flights as a means to guarantee full aircraft, other companies were exclusively charter operations. After 1973, however, increasing fuel prices and competition drove some companies to bankruptcy. As with regional carriers, the history of charter flying is replete with vanished names, including Holidair, Ontario World Air, and Vacationair.

The world has been transformed by air travel (both freight and passenger) and further changes are inevitable. The demise of the Concorde supersonic transport in 2003 demonstrated that technological advances are not always economic.

The evolution of air transport into a mode of travel serving a mass market rather than the elite encouraged the emergence worldwide of low-cost carriers. This, along with the impact of the terrorist attacks of September 11, 2001, forced many of the world's largest and best-known airlines to adapt, seek protection against bankruptcy, or disappear entirely.

The challenges of changing economic and security environments continue to face the air transport industry, while environmental issues and the supply of fossil fuels are growing concerns. Nevertheless, however the players may change, air transport will retain its role as the pre-eminent carrier of people and priority cargo over long distances for the foreseeable future.

Maxwell William Ward (1921–)

MAX WARD had a varied career: Royal Canadian Air Force (RCAF) flying instructor (1941–45), bush pilot (1946–49), house builder (1949–51), charter airline pilot (1951–53), and finally owner and manager of Wardair. Initially, this was a northern enterprise, almost indistinguishable from other bush companies. But Ward moved on from operating de Havilland Canada Otters up to four-engine Douglas DC-6s followed by Boeing 707s, 727s, and 747s, Douglas DC-10s, and Airbus A-300s. Wardair became Canada's largest charter air service and was repeatedly described by travel magazines as the best in the world, with a renowned quality of in-flight customer care. It also operated scheduled airline services from 1984 to 1989, when Wardair was sold to Canadian Airlines International.

Ward received the McKee Trophy in 1973 and in 1975 was appointed an Officer of the Order of Canada. He had a keen interest in history and named many of his airplanes for bush pilots, whose names were thus carried to such destinations as the Caribbean and South Africa. At the height of his success, he purchased a de Havilland DH.83C Fox Moth, the type he had flown in the Northwest Territories between 1947 and 1949. CF-DJB was his personal aircraft until he donated it to the Canada Aviation Museum in 1989.

Canadian
The Saturday Citizen
SPIRIT OF CANADA
1967
THE PLEASURES AND PERILS OF HOT-AIR BALLOONING
DON'T MISS THE
CANADIAN PREMIER
when
Capt. Fearless
CARRIED ABOARD THE SPIRIT OF CANADA
STAN SHELDRAKE
CANADIAN AERONAUT YZB-1
SHELDRAKE
CANADA
Scotchgard FABRIC PROTECTOR
STAN SHELDRAKE
CANADIAN YZB-1 AERONAUT
Spirit of Canada Balloon Associates
Royal Canadian Flying Clubs Association
EXPERIMENTAL AI
WORLD HOT AIR BALLOON CHAMPIONSHIP
FEB. 11-17, 1973
ALBUQUERQUE NEW MEXICO
JOURNEY and AIRCRAFT
LOG BOOK
Price 2.50

RECREATIONAL AVIATION

9

Aviation has touched Canadians in many ways, in war and in peace. Most people are passive consumers, as airline passengers or air-show spectators. For others, however, aircraft are intimate parts of their lives, as tools of their specialist trades or as a means of expressing personal freedom.

facing page: Artifacts and memorabilia of Canada's first modern balloonist, Stanley J. Sheldrake, who constructed and flew his balloon *The Spirit of Canada* during Canada's centennial celebrations in 1967.

Aviation pioneers had been intoxicated with flight as a technical discovery and an adventurous experience. As seen in earlier chapters, barnstormers attempted to support themselves and arouse public interest in flying through appearances at air meets and local fairs. Their success was limited; attention shifted to the next novelty with each new set of newspaper headlines.

Aside from military flying during the First World War, global aeronautical events affected Canadians only sporadically until May 1927. In that month Charles Lindbergh's thirty-three-and-a-half-hour solo flight from New York to Paris was a catalyst. Overnight, the world became "airminded," and an aviation boom resulted. Lindbergh's triumph was all the more spectacular for the failures that had preceded it, including the crash of one contending airplane and the disappearance of another, with five lives lost. The world was fascinated by the man and what he represented. For Canada's sixtieth anniversary of Confederation that year, it was a visiting Lindbergh, not a Canadian, who was the guest of honour in Ottawa, and the pasture where he landed his *Spirit of St. Louis* was

above: The world became "air minded" following Charles Lindbergh's epic trans-Atlantic solo flight in 1927. On a later visit to Ottawa with the *Spirit of St. Louis,* Lindbergh (centre) landed at what is now McDonald Cartier International Airport. Far left is J.A. Wilson, Controller of Civil Aviation, one of Canada's fathers of aviation.

facing page: Robert Bradford's *Armstrong Whitworth Siskin IIIs* portrays the RCAF's famous aerobatic team of the late 1920s and early 1930s, in this instance at the top of a loop from the pilot's perspective. The Siskins were the stars of the Trans-Canada Air Pageant in 1931.

dubbed Lindbergh Field for the next fifteen years. It has since expanded into the present Macdonald–Cartier International Airport.

Lindbergh's example inspired others, including an Anglo-Canadian, to carry out a flight from London, Ontario, to London, England, for a $25,000 prize. The enterprise attracted some two dozen people, one of whom was a notorious con man. It ended with two former Ontario Provincial Air Service pilots, Terrence Tully and James Medcalf, vanishing in the North Atlantic in September 1927. Further attempts were cancelled, and the money went to Tully and Medcalf's widows and children.

Following up on Lindbergh, the global press lavished attention on a succession of pilots, who established a succession of speed and distance records. Errol Boyd, the first Canadian to fly the Atlantic in October 1930, in a Bellanca similar to that held by the Canada Aviation Museum, had been preceded by too many others to make his feat register on any scale of fame. Canada's "stars" were the bush pilots, pioneering frontier flying in the North.

Closely allied to the founding of flying clubs in Canada during the 1920s and 1930s were the great aeronautical tours organized to promote "airmindedness." Small fleets of airplanes, sponsored by manufacturers, local companies, and the Royal Canadian Air Force (RCAF), visited centres across Canada. An air meet in Winnipeg in May 1929 was the first staged in Western Canada. It was followed by an air pageant at Moncton, New Brunswick, in July 1929, in conjunction with the opening of a municipal airport. In 1929 and 1930, American tour groups also included Canadian cities in their itineraries. The tour that had the most impact was the Trans-Canada Air Pageant, staged between July 1 and September 12, 1931. The pageant visited twelve Canadian cities and was seen by over half a million people. The Toronto appearance coincided with a rally-type light airplane race, the Tip Top Aerial Derby, with thirty pilots (chiefly flying-club entrants) participating.

During the 1920s, a number of open-cockpit biplanes of similar design and construction dominated the civil aviation market in North America. Initially, most were powered by war-surplus Curtiss OX-5 engines, then available in large numbers. Seating two passengers side by side in the front seat, with a single rear cockpit for the pilot, a number were

20
20
60
60
59
59

imported into Canada, the most popular being the Alexander Eaglerock, the Laird Swallow, the Waco 9 and 10, and the Travel Air 2000. They were used for flying instruction, barnstorming, private flying, charter trips, and other similar activities. However, only the well-heeled private pilot could afford such aircraft during the biplane's golden age.

In 1925, Geoffrey de Havilland introduced the DH.60 Moth in Britain, the first light aircraft specifically designed for the private pilot. The Moth was widely accepted internationally and was probably the most widely used type of its time. It is often credited as the catalyst that started the flying-club movement. In the late 1920s and 1930s, there were more Moths in Canada than any other type.

Moths found many other uses too. Operating on wheels and floats, they detected forest fires with the Ontario Provincial Air Service and RCAF, and they helped mineral exploration companies find new resources. De Havilland Canada was established in 1928 to assemble, service, and market the Moth; it later became one of Canada's major aircraft manufacturers.

A contemporary of the de Havilland Moth was the Avro Avian, which first flew in 1926. There were two types—the Avro 594 (wooden fuselage) and the Avro 616 (metal fuselage). Wilfrid "Wop" May's famous mercy flight of January 1929 (carrying diphthe-

ria antitoxin from Edmonton to Fort Vermillion, Alberta) used an Avro 594. Most of the Avians built in Canada were acquired by the RCAF and then issued to flying clubs. But the Avian had drawbacks. Its performance was inferior to the Moth's, it was difficult for a pilot to climb into the cockpit while wearing a parachute and, until Avro moved the rudder bars closer to the seat, short-legged students had to pack cushions behind their back so they could reach the pedals.

However, flying was a relatively expensive pastime and the search for a truly affordable light aircraft continued, 1930 being a particularly productive year for prototypes. One of them was the aircraft that is generally credited with triggering the light-plane boom in North America. The single-seat Aeronca C-2 "Razorback" (also known as the "Airknocker" or "Flying Bathtub") had been test flown in October 1929, but its first public appearance was in St. Louis in February 1930. With a two-cylinder engine and only four instruments (oil temperature, oil pressure, altimeter, and tachometer), the selling price was US$1,555; the price was later reduced to $1,245. A two-seater version, the C-3, was also inexpensive at $1,895. Over 160 C-2s and 430 C-3s were manufactured, although fewer than 20 were imported into Canada. Its principal problem was that it was of American manufacture at a time when government assistance to flying clubs was conditional on their using British- or Canadian-made machines.

Although squat in appearance and of modest performance, the C-2s and C-3s were nevertheless easy to fly. They helped make Aeronca one of the leading manufacturers of light aircraft for the private market—and inspired a formidable rival. In September 1930, the Pennsylvania-based Taylor Aircraft Corporation flew its prototype E-2 Cub. It was comparable in price to the Aeronca and some 350 were manufactured by 1936. The E-2 Cub was powered by a 37-horsepower Continental A-40 engine. This was the only cheap, four-cylinder engine available for very small aircraft at this time. The cylinders were horizontally opposed—a configuration that was soon to become the standard of the light-aircraft industry in the United States.

facing page, left: The Avro Avian, pictured here in an ad from *Canadian Aviation,* was a 1920s contemporary of the de Havilland DH.60 Moth. It was built under licence in Canada by the Ottawa Car Manufacturing Company.

facing page, right: A number of open-cockpit biplane designs aimed at the recreational pilot were manufactured in the United States during the 1920s. Most were initially designed around the war-surplus 90 horsepower Curtiss OX-5 engine. All were pleasing in appearance, one of the more popular being the Alexander Eaglerock, advertised in 1929 in *Popular Aviation.*

above: Art Leavens and two passengers with his Waco 9 in 1928. After learning to fly in St. Louis, where one of his instructors was Charles Lindbergh, he and brothers Clare and Walt barnstormed throughout southern Ontario, founding a family business now known as Leavens Aviation.

above: Gertrude de La Vergne, the first licensed female pilot in Alberta, seen here in about 1928.

facing page: Jessica Jarvis featured on the cover of *Star Weekly*, January 6, 1940. She had earned her commercial pilot licence in 1934 but later stated that the licence itself was a farce because she could do nothing with it.

Improvements to the E-2 resulted in the J-2 Cub and, after being acquired by William T. Piper, the company changed its name to Piper Aircraft in 1937. It then produced a new light aircraft, the Piper J-3 Cub, which was subsequently used in such numbers that by some estimates, four out of every five American pilots trained during the Second World War spent some time in one. It was also used as a military liaison and artillery spotter aircraft. As with many types, war-surplus Piper Cubs poured onto the post-war market, and for a time the term "Piper Cub" was synonymous with "light private airplane."

By 1936, other American firms such as Luscombe, Stinson, and Cessna were also manufacturing high-wing, strut-braced light monoplanes. Piper maintained a prominent place in light-aircraft design and manufacture, but over the years it was challenged by Cessna, with its Cessna 150 and Cessna 172, in particular.

As small commercial air companies and private pilots grew more numerous, it became apparent that their interests differed from those of other organizations, such as flying clubs. In December 1952 a proposed tax measure on aviation gasoline spurred a few operators from Ottawa, Kingston, and Toronto to form what became a national organization, the Canadian Owners' and Pilots' Association (COPA).

HOMEBUILTS, HANG-GLIDERS, AND ULTRALIGHTS

Homebuilt aircraft had existed from the beginning of aviation—even Wright biplanes were sold as kits. During the 1930s, a number of homebuilt aircraft designs became available that could be constructed from kits or from available drawings published in magazines such as *Mechanix Illustrated*. Leading types included the Pietenpol Air Camper, Heath Parasol Monoplane, and Corben Baby Ace, which were mainly powered with converted automobile or motorcycle engines.

"Can A Woman Learn to Fly?"

DURING AVIATION'S earliest years, many Canadian men and women confessed to being "air-minded"—in love with the idea of being alone in the sky with the clouds. But in practice, women were restricted to watching from the ground as male pilots thrilled the crowds at public demonstrations.

In 1928, Eileen Vollick became the first Canadian woman to obtain a private pilot's licence. At the age of nineteen, she approached Jack V. Elliot of Hamilton, Ontario, for flying lessons; before agreeing, he waited for consent from the Department of National Defence, which took great pains to investigate her credentials before giving permission. Of her three instructors, two were friendly and one hostile; however, Elliot flaunted her in advertising his flying school. By 1932, more than twenty women across Canada had gained their private pilot's licence.

It soon became evident that female pilots were still having to struggle against the social norms of the time. Despite magnificent exceptions, such as the handful of Canadian women pilots who delivered the wartime trainers, fighters, and bombers from factories to operational units in the United Kingdom for the Air Transport Auxiliary (ATA), women have generally been excluded from most areas of professional flying until relatively recently. Of the five who earned commercial licences before 1940, all encountered difficulties once they tried to move into paying positions, a situation further compounded by the poor economic conditions of the Great Depression.

Typical was the experience of Jessica Jarvis, a member of the Toronto Flying Club who received her private licence in 1931 and was the fourth woman in Canada to qualify for her commercial licence. When war broke out in 1939 she was sure that the Royal Canadian Air Force would accept women but discovered otherwise. Refusing to consider a ground job with the Women's Division, she joined the Navy and spent the war years as a dietician. Realizing that the authorities were generally disinclined to make use of women pilots in wartime—and feeling that there was no place for them in peacetime—she did not renew her membership at the club and never flew again.

CF-AOR
JOHN H MATTHEWS
C-GCGE
The CUB
TAYLOR AIRCRAFT CO.
BRADFORD, PA
JOHN H MATTHEWS

facing page, top: In 1967, the Museum's Aeronca C-2 was restored in the fictitious markings of CF-AOR, the first registered aircraft of its type to be owned and flown in Canada during the 1930s.

facing page, bottom: The Taylor E-2 Cub was the forerunner of the famous Piper Cub. The Museum's example, built in 1935, was eventually restored by a private owner in Toronto before its acquisition by the Museum in 1985.

The Air Camper was designed by an American, Bernard Pietenpol, and his initial design employed a Ford Model A automobile engine. He assembled the prototype with locally purchased wood, fittings fabricated by a blacksmith, and doped bedsheets as covering. The undercarriage was improvised from steel tubing and motorcycle wheels, the propeller was hand-carved from black walnut. Pietenpol spent four years refining the design and devising a single-seat aircraft, the Sky Scout. He then began manufacturing kits that he sold throughout North and South America. Eventually, his company distributed plans only, leading to more than thirty variations as builders substituted lighter engines. Power plants aside, the Pietenpol Air Camper changed little in over seventy years; more than fifty were still flying in Canada in 2004.

Homebuilding activity virtually ceased during the Second World War, but the phenomenon took off—literally—following the war. The Experimental Aircraft Association (EAA) was founded in 1953 by a group of individuals in Wisconsin who were interested in building their own aircraft. Through the decades, the organization expanded its mission to include antiques, classics, vintage warplanes, aerobatic aircraft, ultralights, helicopters, and contemporary manufactured aircraft. Today, the EAA has a worldwide membership, and its week-long annual fly-in held at Oshkosh, Wisconsin, is the world's largest air show. In Canada, many EAA members also belong to the Recreational Aircraft Association, which represents its members in negotiations with Transport Canada (the national aviation regulatory body) regarding amateur-built, classic, and antique aircraft.

Despite their size, homebuilt aircraft have been capable of remarkable feats. In 1978, Robin Morris carried out a non-stop transcontinental flight in a Canadian-designed Zenair CH-300 equipped with autopilot and extra fuel tanks, completing the trip in twenty-two hours, forty-four minutes. Flying from west to east, he had the help of prevailing winds; this also meant that Morris had to make the dangerous trip across the Canadian Rockies when his machine was still heavy with fuel. This aircraft is now in the Canada Aviation Museum collection.

The 1970s saw a new form of flying emerge—hang gliding, which became especially popular in mountainous regions. However, the sport could be dangerous. In the United States, forty-seven hang-glider fliers were killed in 1974 alone, no doubt watched by the spirit of Otto Lilienthal, who died in 1896 in what was essentially an early hang-glider.

facing page: Built in the mid-1950s, the Stits Playboy was the first home-built aircraft of the modern era to be certified in Canada.

In 1976, hang-gliders began to be fitted with engines—not standard aircraft engines, but power plants adapted from chainsaws and snowmobiles. The resulting machines became known as ultralights or microlights. Building largely from available kits, the ultralight aircraft movement spread rapidly. In 1979 the Lazair, a Canadian design manufactured in Port Colbourne, Ontario, was considered the best machine in its class at an Experimental Aircraft Association fly-in at Lakeland, Florida.

Ultralight flying was encouraged by economics and technology. Although more people wanted to fly, traditional light aircraft remained expensive and insurance costs were increasing. At the same time, epoxy, fibreglass, foam, and light synthetic fabrics reduced weight, costs, and complexity. Ultralights proliferated most rapidly in the United States where, defined by weight and load capacity, they were regarded as little more than powered hang-gliders, and neither aircraft nor pilots were licensed. It was only in 1982 that the Federal Aviation Administration (FAA) laid down its first rules governing ultralights. However, it defined the craft in very restricted terms—powered ultralights could not exceed 115 kilograms (253 pounds), could carry no more than 19 litres (5 U.S. gallons) of fuel, and could have a stall speed of no more than 45 kilometres per hour (28 miles per hour).

In 1983, Canadian legislation required ultralight aircraft to be registered and ultralight pilots to be trained and licensed. In Canada today, ultralights may be flown with an Ultralight Pilot Permit, a Recreational Pilot Permit, or a Private Pilot Licence.

GLIDING AND BALLOONING

Gliding as a sport has had a sporadic development. During the 1920s, the first priority of those "selling" aviation was to build a light airplane that was cheap and easy to fly. Aircraft like the de Havilland Moth met those standards, but glider tugs (small aircraft capable of towing a glider into the air) trailed a few years behind. Often gliders were launched by an automobile tow. The first time a glider was towed by a powered aircraft in Canada occurred in Vancouver in July 1930. Thereafter, gliding and soaring clubs formed, functioned, and vanished in haphazard ways. All surviving gliding associations ceased operations upon the outbreak of the Second World War, but by 1944 some were preparing to resurrect themselves. The Montreal Soaring Council, for example, was formed in 1944 to

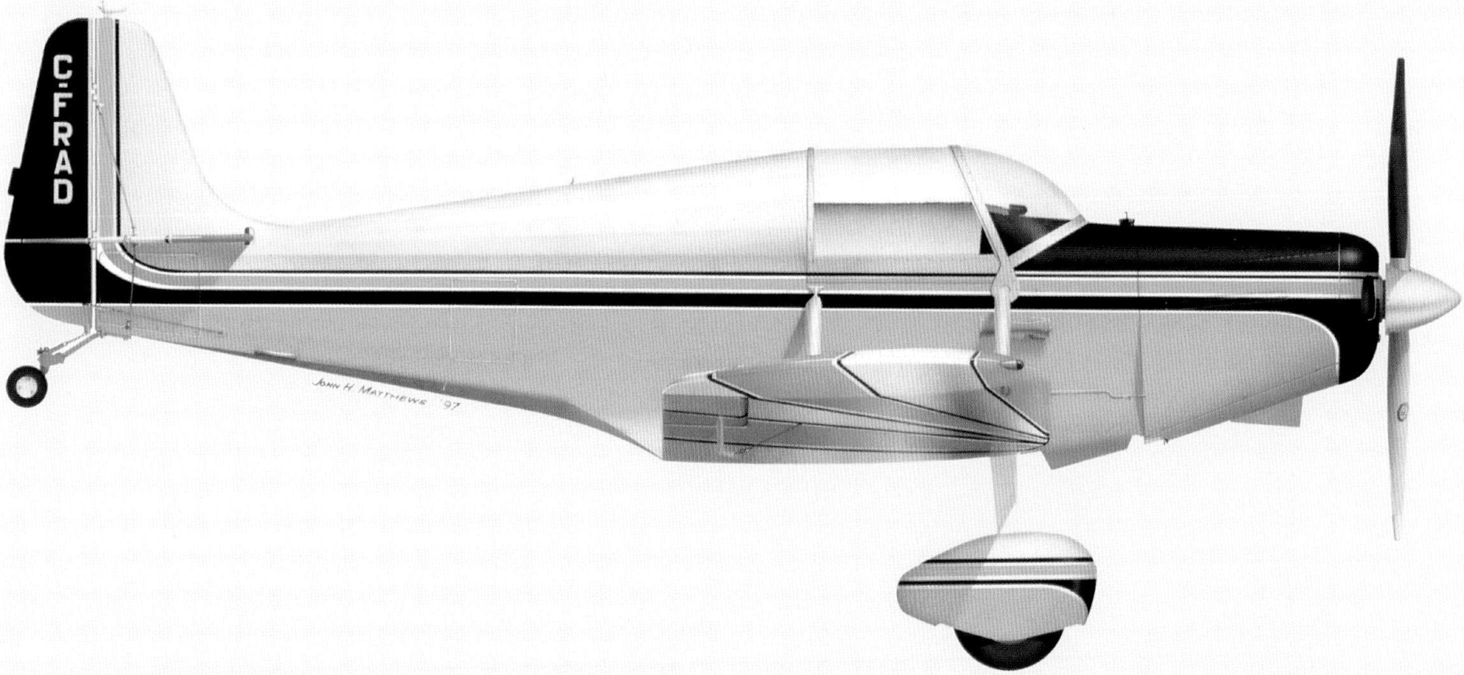

Keith Hopkinson
and the Stits SA-3A Playboy

KEITH HOPKINSON of Goderich, Ontario, was the first to construct a homebuilt aircraft in Canada in the post-war period. He also promoted homebuilt aircraft vigorously, emphasizing their engineering standards and high degree of workmanship, and is credited with persuading the Department of Transport to introduce a special licensing system for them, thus freeing their builders from the regulations that governed the construction of commercially manufactured machines. (To be classified as a homebuilt, 51 per cent of the aircraft must be built by the constructor.)

In 1955, Hopkinson assembled a Stits SA-3A Playboy from a kit. He modified it extensively, using the nose cowl from a Piper J-3 Cub, the propeller spinner from a Cessna 170, wing struts from a de Havilland Tiger Moth, landing gear from a Cessna 140, and wheel fairings from a Stinson 108. It was the first modern example of a homebuilt produced in Canada. After he died, the Playboy was sold, and it was later purchased by the Museum in 1978.

above: The Corben Super Ace, an American design of the 1930s available to the amateur builder, is pictured in 1935 on the front cover of a leading aviation magazine.

facing page: A Piper J-3 Cub undergoes maintenance at Curtiss-Reid Flying Services in Montréal after the Second World War while another is readied for flight in the hangar. The term "Piper Cub" was once synonymous with "light private airplane."

co-ordinate activities of several clubs, although it did not begin to function until 1946. Meanwhile, the Soaring Association of Canada had been founded in 1945. Its first step was the establishment of a gliding school at Carp, Ontario. The first Canadian gliding and soaring competitions were held at Downsview, Ontario, in September 1946.

Gliding and soaring differ in that gliding usually involves shorter flights to a planned landing site, whereas soaring employs more efficient sailplanes that harness updrafts and rising thermal columns of air to attain higher altitudes and longer distances. The two-seat Harbinger was a sailplane designed in Toronto by Waclaw Czerwinski and Beverley S. Shenstone for a 1947 design competition sponsored by the British Gliding Association. The first of the type was built in England and flown in 1958. A modified design was built in Toronto, beginning in 1949 but, after many delays, not being completed until 1975. After a brief flying career of thirty flights, including one of three hours, it was retired and is now in the Museum's collection.

One problem faced by soaring clubs after the war was that urban airspace was becoming increasingly busy, forcing members, tugs, and gliders to seek pastoral sites such as Hawkesbury and Pendleton, Ontario, which became the homes of the Montreal Soaring Council and the Gatineau Gliding Club respectively. The Soaring Association of Canada (SAC) received assistance from firms such as Goodyear, Canadian Pratt and Whitney, and Massey-Harris. Growth was steady and by 1969 SAC had 1,022 members. By 2003, there were 1,250 members organized into thirty-three clubs across the country. That year there were 674 gliders and sailplanes registered in Canada.

Balloon technology has fostered a new sport. Eighteenth-century balloons had been borne aloft with hot air produced from burning straw (which apart from being dangerous was suitable only for short flights). Today, self-contained propane tanks, burners, and lightweight polyester have combined to produce the modern aerostat, but the traditional champagne toast has been retained to celebrate a passenger's first flight.

CURTISS-REID FLYING SERVICE LTD.

above: The name of the Museum's Harbinger sailplane, designed by Waclaw Czerwinski and Beverley Shenstone, suggested the beginning of a new era in two-seat sailplane design when it was entered in a 1947 design competition in the United Kingdom.

facing page: Carl Hiebert at the Canada Aviation Museum during his epic "Gift of Wings" cross-Canada flight in 1986. The Museum later acquired his Spectrum Beaver ultralight, complete with inflatable floats.

In 1964 Stanley J. Sheldrake, a sport parachutist from Beamsville, Ontario, became interested in ballooning. He assembled CF-VOZ, named *Spirit of Canada,* a spectacular red-and-white craft that he completed early in 1967. As one of Canada's first modern balloonists, Sheldrake made numerous ascents at Canadian centennial celebrations.

Centennial celebrations also included ascents by American and European balloonists, and the Alberta Free Balloonists' Society, formed in 1967, sponsored flights across the West. By the mid-1970s, there were some 35 registered balloons in Canada, and the first Canadian National Hot Air Balloon Championship was staged at Grande Prairie, Alberta, in 1979. Since then, hot air balloons have become familiar objects in Canadian skies, appearing in large numbers at balloon festivals across the country. As of 2003, there were 450 registered hot air balloons in Canada, flown by 270 active balloon pilots.

FROM EAGER YOUTHS to experienced senior citizens, thousands of Canadians actively fly as recreational pilots each year, enjoying the experience as participants rather than consumers. For some it brings physical enjoyment; for others it approaches a religious event. All can harken back to the sense of joyous peacefulness conveyed by John G. Magee's poem "High Flight," composed in very different circumstances in 1941 and resonating still to this day.

Ultralight Heroes

CANADIAN ULTRALIGHT pilot Bill Lishman has achieved fame for teaching Canada geese to migrate south, using the process of imprinting discovered in the 1950s by Nobel laureate Konrad Lorenz. Lishman began experimenting in 1986 to determine whether he could compel newly hatched goslings to "adopt" his ultralight aircraft as their "mother." Having trained the geese to fly behind his machine, he undertook to lead a flock on a migratory flight from Ontario to Virginia in 1993, with the aim of eventually doing the same for endangered species such as the whooping crane. This inspired a book (*Father Goose*) and a movie (*Fly Away Home*) as well as further studies of bird migration.

Carl Hiebert, like Lishman, began as a hang-glider pilot, but an unfortunate accident put him in a wheelchair in 1981. He took up ultralight flying, a sport that eventually inspired him to write: "To look down and see an empty wheelchair was the best therapy that could have happened." Within two years, Hiebert had started his own flight school as the first Canadian paraplegic flying instructor. In 1986, he flew across Canada, raising $100,000 for the Canadian Paraplegic Association and landing at Expo 86 in Vancouver. He soon gained international recognition as a pilot, photographer, and author whose spell-binding books blend his love of country with deep spiritual conviction.

The Canada Aviation Museum's Spectrum Beaver, acquired in 1987, was flown by Hiebert on his epic transcontinental journey. This Canadian design is powered by a 48-horsepower, two-cylinder Rotax engine and is fitted with inflatable floats.

MAPLE LEAF
RW BRADFORD

TRAINING 10

MANY PIONEER AIRMEN were both designers and self-taught test pilots. In Europe and North America, the process of teaching others to fly began in 1909 at a personal level; the first specialized flying schools (civil and military) were formed about 1910. Often a manufacturer, such as Bristol, Curtiss, or Blériot, established a flying school in conjunction with the factory making the airplanes. This practice continued for decades.

Initially, few aircraft were designed specifically for training. During the First World War, types that had become obsolete in combat were pressed into service as trainers. Thus, we find early Avro 504 aircraft conducting bombing raids in 1914; two years later the Avros no longer flew in battle but were ubiquitous in British flying schools. Similarly, the Maurice Farman Série 7, dubbed the "Longhorn" because of its forward-projecting elevators, and its successor, the more orthodoxly shaped Série 11 "Shorthorn," were both flown as scouts in 1914–15 by France, Britain, Belgium, Australia, Italy, and Russia. By the end of 1915, most had been withdrawn from combat and switched to instructional roles. Their long skids, which prevented the aircraft from nosing over when landing, were features shared with other trainers, including the Avro 504. Shorthorns continued in this role until 1918.

Early training taught the basic elements of flying but emphasized dangers to be avoided, such as stalls and spins. By 1916, the dynamics of flight controls were more

facing page: Curtiss "F" Flying Boat: The Maple Leaf, by Robert Bradford. The Curtiss Aviation School at Long Branch, Toronto, was one of Canada's earliest flying schools. Run by Curtiss Aeroplanes & Motors Limited, the school trained pilots destined for the Royal Naval Air Service during 1915 and 1916.

below: During World War I, aircraft types that had become obsolete for combat were pressed into service as trainers. The Museum's Maurice Farman Shorthorn was one of four sent to Australia for flight training in 1917.

facing page: This student lecture in aircraft structures at No. 4 School of Aeronautics, University of Toronto, was photographed during the Royal Flying Corps (Canada) training scheme in 1917.

fully understood and these manoeuvres were practised routinely. Major Robert R. Smith-Barry, Royal Flying Corps, developed the "Gosport" system of instruction (so named for the school where he taught, in which more emphasis was placed on the theory of flight and aerobatics). The airplane was to be seen not as a threatening mount but as an even-tempered, reasonable machine. Smith-Barry also stressed the need for a standard curriculum and for continuing instruction after students had made their first solo flight rather than leaving them to themselves.

THE FIRST WORLD WAR AND AFTER

The first two successful Canadian flying schools were the Curtiss Aviation School in Toronto and the British Columbia Aviation School in Vancouver, both organized in response to the outbreak of war. The latter operated for one year in 1915 and produced two graduates, whereas the Curtiss School lasted two seasons, 1915 and 1916. A former student recalled that the only "instruction" he received was the injunction shouted in his ear by his instructor, "Steer, damn you, steer."

The Royal Flying Corps (RFC, later amalgamated into the Royal Air Force, RAF), which was expanding and suffering increased casualties in 1916, established an ambitious training program in Canada. It was associated with Canadian Aeroplanes, the Curtiss JN-4 (Can.) trainer aircraft factory, and it recruited and instructed in many trades, including pilots, gunners, and mechanics. By war's end, the organization occupied quarters at Hamilton, Toronto, Long Branch, Beamsville, Armour Heights, Leaside, Camp Rathbun (Deseronto), Camp Mohawk (Deseronto), and Camp Borden. The quarters that they occupied included former schools, a prison, and much of the University of Toronto. The staff included about twelve hundred women, who had been recruited to be mechanics and drivers.

The RFC (Canada) training was very sophisticated, incorporating aviation medi-

cine and the psychological screening of candidates. It also followed the modern Gosport system and even used primitive flight simulators. Some aircraft were adapted to skis, enabling the first sustained winter flying in Canada, and special clothing was developed for cold-weather training. A typical student went solo after seven hours of instruction and spent five months in Canada before going overseas.

The RFC/RAF (Canada) training scheme enrolled 9,200 cadets in 1917–18. Of these, 3,135 completed pilot training and more than 2,500 were sent overseas. In addition, 137 observers were graduated, of whom 85 were sent overseas. The program also turned out at least 7,400 mechanics. Probably the most famous graduate was Lieutenant Alan A. McLeod, VC, who trained at Long Branch and Camp Borden and received his wings in July 1917. Other distinguished alumni included fighter aces Captains Donald R. MacLaren and William G. Claxton.

The training scheme eventually included many Canadians at all levels. The Department of Militia and Defence assigned paymasters, doctors, and other non-flying personnel

The Curtiss JN-4 Canuck

EARLY FIRST WORLD WAR training aircraft tended to be machines that had been deemed obsolete for front-line service. Organized training speeded the evolution of specialized training aircraft such as the Curtiss JN-4 (Can.). Based on an American design, it was modified by Canadian Aeroplanes Limited to meet Royal Flying Corps (RFC) training needs, notably by removal of the wheel control and substitution of a joystick. Some JN-4s were adapted to accommodate camera guns, reconnaissance cameras, and machine guns. The JN-4s flown in Canada sported a variety of colourful and distinctive markings, including maple leaves, terriers, black cats, shamrocks, and Jolly Roger insignia. Some were named for cities such as Edmonton and Montréal; at least six bore names commemorating battles of the War of 1812. Had the war continued, JN-4 production would have been superseded by construction of the more advanced Avro 504 (Canadian).

More than 1,200 JN-4 (Can.) machines were built, of which 680 were exported to the United States following its 1917 entry into the war. Pilots enjoyed flying the JN-4, but with a 90-horsepower Curtiss OX-5 engine, it was slightly under-powered. It had a maximum speed of 120 kilometres per hour (75 miles per hour), cruised at 100 kilometres per hour (60 miles per hour), and had a ceiling of 3350 metres (11,000 feet). Cheaply available in large numbers after the war, the Canuck was a favourite of barnstormers during the early 1920s.

to the various schools and headquarters. Increasingly, Canadian pilots and observers joined the instructional staff. Some were recent graduates of the scheme; others were veterans of the Western Front. By November 1918, Canadians commanded the School of Aerial Fighting at Beamsville; two of the three training wings, twelve of the sixteen training squadrons, and roughly 60 per cent of all instructors were Canadians.

The RFC/RAF (Canada) scheme was the single most powerful influence in bringing the air age to Canada. The pilots and mechanics trained were a foundation on which was built Canadian commercial aviation, including bush flying, and the Royal Canadian Air Force.

Immediately following the First World War, a few enterprises trying to find a commercial niche, such as the Jack V. Elliott Company, listed flying lessons among their activities. The Canadian Air Force/Royal Canadian Air Force (CAF/RCAF) gave refresher flying training to wartime veterans, many of whom returned to civil life and struggling companies. In 1923, the RCAF commenced training new pilots. However, neither the air force nor the scattered schools produced sufficient aircrew for even the nascent Canadian aerial market.

facing page: A Curtiss JN-4 (Can.) trainer adapted to accommodate a forward-firing Lewis machine gun. As the gun is not synchronized to fire through the spinning propeller, it must fire over it. This photo was most likely taken at the School of Aerial Fighting in Beamsville, Ontario, in about 1918.

above: The perils of flight training are not lost on the flight cadets marching past this accident scene at Camp Borden in 1917.

below: The appearance of the de Havilland DH.60 Moth in Canada coincided with the rise of the flying clubs. In the mid-1930s, Dr. and Mrs. J.J. Green of the Ottawa Flying Club pose with a privately owned machine assembled by de Havilland Canada in 1930.

The appearance in Canada of the de Havilland DH.60 Moth in 1927 coincided with the rise of flying clubs offering instruction outside the circles of military or commercial schools. The Toronto Flying Club began operations on May 1, 1928, closely followed by others across the nation. In November 1929 the Canadian Flying Clubs Association was formed to co-ordinate activities and to represent the clubs in dealings with the federal government.

The clubs were not entirely private but were supported by the Department of National Defence, which provided each new club with two aircraft and then kept a watchful eye on the standards of maintenance and training. The curriculum was thorough and standardized. Training on the ground included subjects such as theory of flight, rigging, compasses, charts, course plotting, engine components, meteorology, and aids used in cross-country flying—all of which climaxed with a written examination. Actual flying instruction was anywhere from forty to sixty hours, often in short flights of ten or fifteen minutes. Even then a student could fail if he or she was not knowledgeable about air regulations. Assuming the club had a qualified instructor, the new pilots could go on to earn commercial licences.

facing page: Curtiss JN-4 (Can.) & JN-4A, by Robert Bradford, depicts a typical working day at the No. 1 Aerial Fighting School, Beamsville, Ontario, in 1918. The American-designed Curtiss JN-4A (upper centre) and the Canadian-built JN-4s (Can.) shown elsewhere were structurally different, most obviously in the rudder shapes.

Flying clubs did more than offer instruction. They drew community attention to aviation in many ways, from dropping pamphlets and advertising to participation in local air shows.

The flying club movement started quickly but was battered by the Depression, when its activity declined. Instructors and mechanics sometimes went without pay for months, and recovery was slow. Nevertheless, the RCAF regarded the clubs as a reserve force, and when Auxiliary squadrons were formed in the 1930s their membership often overlapped that of the local flying associations. The benefit to the RCAF is attested to by the many club graduates who later joined the force.

c562
c1311
c1308

This de Havilland DH.60X Cirrus Moth was built in 1928 and was restored by volunteer de Havilland Canada employees in the early 1960s after its donation to the Museum.

This relationship became more intimate upon the outbreak of the Second World War, when twenty-two flying clubs were reconstituted as Elementary Flying Training Schools or created spin-off schools as part of the British Commonwealth Air Training Plan (BCATP). Three other clubs organized Air Observer Schools, which trained navigators. The wartime role of the Canadian Flying Clubs Association was recognized in 1944 when it had the prefix "Royal" bestowed upon it. Following the war, the clubs snapped up surplus aircraft at nominal prices. The Winnipeg club, for example, acquired eleven de Havilland Tiger Moth elementary trainers at $300 each and two Cessna Crane advanced trainers at $600 each.

Eventually the war-surplus aircraft had to be replaced. The de Havilland Canada Chipmunk, much favoured by the RCAF, was too expensive for the clubs, which also preferred side-by-side seating arrangements for students and instructors. The diversity of American light aircraft available from such firms as Aeronca, Beechcraft, Cessna, and Piper ensured that no single type would ever again monopolize private flight training as thoroughly as the DH.60 Moth had in its prime.

THE BRITISH COMMONWEALTH AIR TRAINING PLAN

The largest single training program ever conducted in Canada was the British Commonwealth Air Training Plan (BCATP), organized to train aircrew from throughout the British Empire and Commonwealth for the Second World War; it operated from 1940 to

1945. The BCATP was considered Canada's greatest contribution to the ultimate Allied victory in the air overseas, the nation itself being described as "the Aerodrome of Democracy" in a letter from President Franklin D. Roosevelt to Prime Minister Mackenzie King (though in fact the letter had been drafted by Lester B. Pearson, then on diplomatic assignment to Washington, D.C.).

Thousands of young men experienced the personal triumph of their first solo flights (usually after about 10 hours of instruction), followed months later by receipt of their flying badges at Wings Parades (having by then flown roughly 140 hours). These ceremonies were enshrined in the film *Captains of the Clouds* (1942), during which Air Marshal W.A. "Billy" Bishop, playing himself, pinned wings on real students at No. 2 Service Flying Training School in Ottawa.

On the ground, Canadians became more conscious of both the aircraft and the men who flew them. Yellow training planes—so painted to make them more visible and to prevent collisions—were seen daily. In cities and small towns, civilians met successive waves of Commonwealth and European trainees, many with odd accents and from strange-sounding hometowns. In Ontario, the exiled Norwegian government ran its own training system, known as "Little Norway," which dovetailed with the BCATP.

No training scheme could hope to mimic the wartime conditions that its graduates would encounter overseas. The BCATP was no exception. Away from the Atlantic and Pacific coasts, blackouts were unknown. Many navigation problems were resolved by using a convenient railway line (the "iron compass") or by reading the names of towns on

The John C. Webster Memorial Trophy

After its founding, the Canadian Flying Clubs Association assumed responsibility for administering the John C. Webster Memorial Trophy, established in 1932 in memory of the first Canadian to compete in Britain's prestigious King's Cup Air Race (a rally-type competition). It recognized the most skilled private pilot in Canada, based on flying, navigational skills, knowledge of regulations, and ability to draft flight plans. The trophy, designed by renowned Canadian sculptor Robert Tait McKenzie, depicts Icarus and thus represents both flight and youth.

The Webster Memorial Trophy competitions were held annually from 1932 to 1939, suspended during the Second World War, resumed in 1946, and discontinued again after 1954 owing to administrative costs. They were revived again in 1980 with support from Air Canada and are supervised by the Canadian Sport Aeroplane Association. The trophy is housed in the Canada Aviation Museum.

this page, from top to bottom: Vital to the British Commonwealth Air Training Plan (BCATP) was the DH.82C Tiger Moth primary trainer (top), which equipped many Elementary Flying Training Schools in Canada. A counterpart of the Tiger Moth was the Fleet Finch (centre). More than 400 of these sturdy basic trainers served the RCAF in the early stages of the BCATP. In mid-1942, the RCAF began replacing Tiger Moths and Finches with the Fairchild Cornell (bottom). A more complicated primary trainer was needed to facilitate the student's transition to more advanced training types, such as the North American Harvard.

facing page, from top to bottom: The North American Harvard (top), acquired by the RCAF in 1940, was its most famous advanced trainer for a quarter of a century. The Avro Anson V (centre), used at all the navigator-training schools in the BCATP, was liked for its roominess and comfort. It featured a moulded plywood fuselage, and a number were flown after the war by civilian operators on photographic or electromagnetic surveys. A British design of the early 1930s, the Fairey Battle (bottom) was obsolete by the beginning of the Second World War. Most served on training duties and many were shipped to Canada, where they were widely used as gunnery trainers and target tugs.

The colour profiles on these pages portray aircraft actually in the Museum's collection.

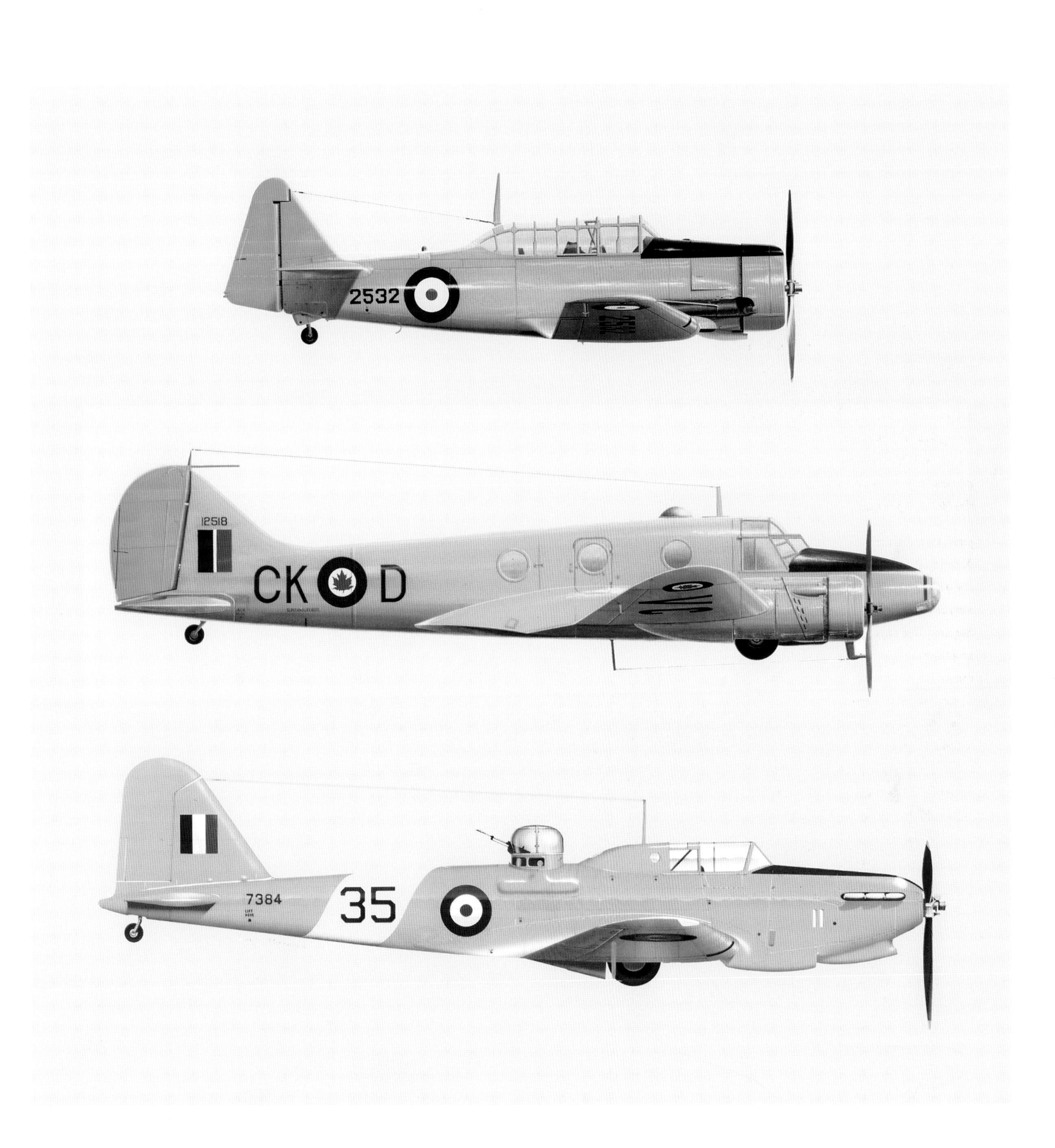
2532
12518
CK
D
7384
35

above: A typical Elementary Flying Training School flight line, as students and instructors walk to their Tiger Moths in the early morning at No. 20 EFTS, Oshawa, Ontario. Nearly 50,000 pilots were trained in Canada under the British Commonwealth Air Training Plan.

facing page: Murton Seymour at the controls of a Curtiss biplane. He was trained to fly by William Stark, Canada's second licensed pilot, in 1915.

Prairie grain elevators. On the other hand, there were few discomforts to match winter flying in an Avro Anson—which had been designed for Britain's milder climate—or in an open-cockpit Stearman PT-17 Kaydet.

A training program such as the BCATP required more than ordinary flying schools. Wartime specialist trades such as bomb aimers demanded their own centres; bombing and gunnery schools were located near lakes, where flying was safer and gunfire accuracy could be assessed by bullet splashes. There were also special schools to teach instructors. Ultimately, a team of inspectors from the Central Flying School in Trenton, Ontario, toured the country to assess instructor performance and curriculum relevance. The system was backed up by thousands of mechanics, produced in their turn at the School of Technical Training in St. Thomas, Ontario.

The Second World War witnessed many biplane primary trainers (such as the de Havilland Tiger Moth, the Fleet Finch, and the Stearman Kaydet) but also monoplane primary trainers (like the Fairchild Cornell and the Miles Magister). The prevalence of monoplanes, whether in military or commercial roles, spelled the long-term demise of biplane instructional aircraft, but sentiment and the magic of the "wind in the wires" ensured the preservation of many older machines.

As aircraft and their roles became more complex, air forces recognized the need for instructional machines that were more advanced than the basic elementary trainers, yet not as formidable as combat aircraft. The North American Harvard (U.S.) and the Miles Master (Britain) were designed to fill just this need in single-engine aircraft; twin-engine advanced trainers were developed chiefly from light transports, producing the Avro Anson and Airspeed Oxford. Some were failed or retired combat aircraft, such as the Fairey Battles and Bristol Bolingbrokes that were recycled at BCATP schools.

Murton A. Seymour (1892–1976)

Murton Seymour, a lawyer by profession, never lost his interest in aviation after being trained to fly in 1915 by William Stark, Canada's second licensed pilot. Following overseas service with the Royal Flying Corps during the First World War, Seymour returned to Canada in 1917 as a staff officer with the Royal Flying Corps (Canada) flying training scheme, where he played a role in organizing its training program.

After the war, he became a key figure in the flying clubs movement of the late 1920s and helped establish the Canadian Flying Clubs Association. As its president in 1939, and convinced that the gathering war clouds meant trouble for the Commonwealth, he approached the government with a view to expanding the activities of the flying clubs for defence purposes. An agreement was reached that saw the clubs take on the task of elementary flying training within the British Commonwealth Air Training Plan (BCATP). More than forty thousand air force personnel received their first instruction at these elementary training schools.

To Seymour, the syllabus of these later years must have seemed a far cry from the initial training he received from Stark in 1915. This involved sitting on the leading edge of the lower wing of Stark's Curtiss Pusher and watching the actions of his instructor during demonstration flights. Then he practiced taxiing on the ground with Stark sitting on the wing beside him shouting instructions. A block was placed under the foot throttle to control the amount he could depress the pedal, limiting the engine's power. As Seymour became more proficient at taxiing, the block was shaved down until finally he could generate enough power to make short hops, about 1 metre (3 to 4 feet) off the ground. After that accomplishment, he was on his own.

Student Pilot, RCAF, Elementary Flying Training School:
They allowed ten hours for solo…
If you hadn't soloed by ten, most people just washed out,
and that was the end of your pilot career.
Élève-pilote, ARC, École élémentaire de pilotage
Au bout de dix heures d'instruction en vol, les élèves
devaient être capables de voler pour la première fois en solo.
…si, après dix heures, on n'avait toujours pas
la permission de voler seul, on était fini,
et c'est là que se terminait notre carrière de pilote.
MOTH
RCAF
20
EFTS
GALT AIRCRAFT SCHOOL
Chief Instructor
E
RCAF
WAG
RCAF
MPH
120 110 100 90 80 70 60 50 45

JET TRAINING AND SIMULATORS

The RFC/RAF (Canada) training scheme of 1917–18 and the BCATP of 1940–45 were succeeded during the Cold War by a program in which the RCAF hosted training of aircrew from North Atlantic Treaty Organization (NATO) countries. The original scheme ran from 1950 to 1957 and graduated 4,600 pilots and navigators from ten countries. It was followed by arrangements in which Canada continued to offer flying training to some NATO pupils (particularly from Denmark, the Netherlands, Norway, and West Germany) as well as to a trickle of students from newly independent African and Asian countries.

"Privatization" concepts affected military flight training, and in 1998 a new program was instituted. NATO Flying Training in Canada was a partnership of the Canadian Forces, Bombardier Aerospace Corporation, and participating nations, which also included Singapore (not a NATO nation) and Hungary (once considered a hostile country). Conducted at Portage la Prairie and Moose Jaw, Saskatchewan, and Cold Lake, Alberta, the program trained students in primary flight training up to modern jet fighter standards. The principal elementary training airplane, the Raytheon CT-156, was named the Harvard II in tribute to an earlier generation of pupils and instructors. Meanwhile, NATO air forces have used Canadian airspace in Labrador and Alberta to conduct battle combat training.

Jet technology led to specialized jet training aircraft. Curiously, advanced jet trainers appeared before *ab initio* trainers designed for beginners. This was due to certain combat aircraft being developed from single-seat to dual-control versions. The de Havilland Vampire and Gloster Meteor, which began as single-seat fighters, appeared in trainer modes; the Lockheed P-80 Shooting Star gave rise to the T-33 Silver Star, designated CT-133 in the RCAF and integrated Canadian Armed Forces. No training type served in the Canadian forces longer than the Canadair-built CT-133, which entered service in 1953 and was retired in 2005. Apart from pilot training, it served in the role of target tug; even experienced pilots routinely practised instrument flying on it. Although its role in pilot training declined from 1974 onward, the CT-133 was pressed into other instructional tasks such as representing the enemy in air exercises, using electronic jamming gear to challenge Canadian airborne and ground-based defences.

Ab initio (elementary) jet trainers appeared in 1952 with the French Fouga CM.170 Magister. Britain and the United States followed in 1954. Canada's contribution to this

facing page: Artifacts and memorabilia associated with the British Commonwealth Air Training Plan. The leather flying helmet (upper right) has a Gosport speaking tube attached, which allowed the instructor to communicate with the student while in the air. The mechanical air speed indicator (lower left) was once mounted on the wing strut of a primary trainer.

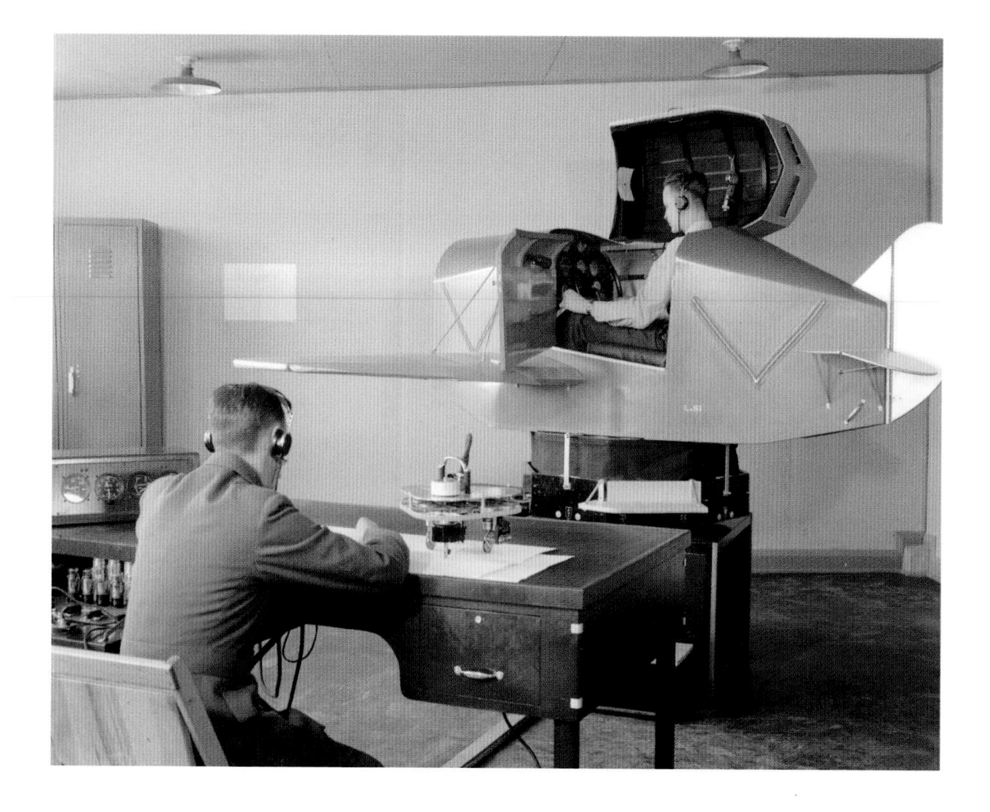

Flight student seated in a Link trainer. An automatic course recorder can be seen on the instructor's desk. Known as the navigational "crab," this device enabled the instructor to track the student's progress over a predetermined course.

genre, the Canadair CT-114 Tutor, first flew in 1960. It has since become familiar to Canadians through the aerobatic teams that have entertained crowds at air shows and special events. The Golden Centennaires were featured during the 1967 centennial; they have since been replaced by the Snowbirds. However, the Tutor's principal task has been pilot training, where its side-by-side seating is especially useful.

During the First World War, various devices were built to teach aerial gunnery and bomb-dropping without leaving the ground; student pilots were familiarized with controls while seated in basic, wingless airframes. However, the true prototype of the flight simulator was the Link Trainer, developed in 1930 by Edwin Link. His experience with both building church organs and flying led him to design a mock airplane that used compressed air to simulate climbing, diving, and banking within limited parameters. Although air forces were beginning to examine the possibilities of all-weather flying, Link found few buyers of his device other than amusement parks.

By 1935, however, official interest was quickening, particularly in "blind" instrument flying and radio-range navigation. At the same time, Link had created the Automatic Course Recorder, which enabled an instructor outside the device to monitor and track the progress of the student inside. This feature allowed mistakes to be recorded and later corrected. The RCAF and Royal Air Force (RAF) bought their first Link Trainers in 1937; Trans-Canada Air Lines acquired one in 1938.

Link Trainers were present throughout wartime air force training, from initial aircrew selection at Manning Depots to instrument-flying practice at overseas units. There were cases of wartime navigators and bomb aimers with only basic Link experience who flew damaged bombers back to England after the pilots had been killed or wounded.

The concept of simulator training has expanded greatly ever since. Canadian Aviation Electronics (CAE), founded in Ottawa in 1947 with eighteen employees, ini-

Airmen Trained in the BCATP (1940–1945)

	RCAF (Canada)	RAF (Britain)	RAAF (Australia)	RNZAF (New Zealand)	TOTAL
Pilots	25,747	17,796	4,045	2,220	49,808
Navigators	12,855	13,882	1,643	1,583	29,963
Bomb Aimers	6,659	7,581	799	634	15,673
Wireless Air Gunners	12,744	755	2,875	2,122	18,496
Naval Air Gunners	–	704	–	–	704
Air Gunners	12,917	1,392	244	443	14,996
Flight Engineers	1,913	–	–	–	1,913
TOTAL	72,835	42,110	9,606	7,002	131,553

tially produced machines little more advanced than the wartime Links. In 1951, they designed an advanced simulator for Avro Canada CF-100 jet fighter training. Since that time, the company has grown to six thousand people (including twelve hundred in the United States) engaged in designing, manufacturing, and operating simulators of craft from helicopters to larger transports and combat aircraft. By 2003, CAE had sold approximately 440 simulators to airlines and armed forces and operates training facilities around the world. These devices can be "flown" twenty hours a day, seven days a week, without the costs associated with large aircraft. Pilots experience and learn emergency procedures that cannot be practised safely in real conditions, such as how to deal with wind shear and engine fire. The modern simulator instructor has up to five hundred "malfunctions" available to provide in-depth training of an aircraft's abnormal or emergency systems and procedures. With their hydraulically generated movements and high-tech visual systems, the machines are so sophisticated that pilots proficient on one aircraft can be completely trained on the simulator for a new type before ever flying the aircraft itself.

Today, there is scarcely a modern airport that does not offer some degree of pilot instruction. Flight training has expanded from a coterie of heroic amateurs to a global network of schooling. Although such work draws upon the most advanced technologies, it must inevitably be combined with human teaching skills. It is also a profession in which mentors and pupils are united by their mutual dedication to flight itself.

RW BRADFORD

AIRCRAFT MANUFACTURING *in* CANADA

THE AERIAL EXPERIMENT ASSOCIATION (AEA) was dissolved shortly after Canada had formally entered the aerial age with the successful flight of the *Silver Dart* in 1909. Alexander Graham Bell encouraged the two Canadian members, J.A.D. McCurdy and F.W. "Casey" Baldwin, to continue the association's work as a commercial venture, with Bell providing financing and facilities. McCurdy and Baldwin formed the Canadian Aerodrome Company at Baddeck, Nova Scotia, and immediately began design and construction of a biplane based on the *Silver Dart*. The new aircraft, named Baddeck No. 1, was completed in July 1909, the first powered aircraft designed and manufactured in Canada. This aircraft and the *Silver Dart* were demonstrated to Canadian military authorities at Camp Petawawa, Ontario, in August 1909, but both aircraft were badly damaged in accidents during preliminary trials. This confirmed the preconceived poor opinions held by the military establishment and they took no further interest in the young company's products. As no substantial Canadian civil or military market for aircraft existed, the Canadian Aerodrome Company disappeared from the scene, closing the first chapter in the story of Canadian aircraft manufacturing.

During the next few years, a few Canadians attempted to build and fly aircraft with varying degrees of success, but there were no further attempts at aircraft manufacturing. Interest in aviation was high; pioneer flyers were popular attractions at exhibitions

facing page: Curtiss Canada Bomber, by Robert Bradford. This design was the first twin-engined aircraft built in Canada. The prototype is shown flying over the Curtiss Aviation School at Long Branch, Ontario, in 1915.

The Baddeck No. 2, one of three aircraft designed and built by the Canadian Aerodrome Company. A monoplane built by the company for a private citizen in the U.S. became the first Canadian aircraft manufactured for export.

and fairs across the country, but there was no domestic market large enough to support manufacturing even on a limited basis.

CANADIAN PRODUCTION UP TO 1945

With the outbreak of the First World War in 1914, the picture gradually began to change. In Europe, aircraft were accepted as useful instruments of war and there was a demand for aircraft and qualified aircrew. First on the scene in Canada was a short-lived Montréal concern, the Canadian Aircraft Works, which built one or possibly two aircraft. However, no official interest was shown and it soon faded away.

Late in 1914, a more successful undertaking began. McCurdy, representing the Curtiss Aeroplane Company of Hammondsport, New York, approached the Canadian government with a proposal to manufacture Curtiss aircraft in Canada and train pilots for the armed forces. Although the government expressed little interest, McCurdy continued efforts to interest both Canadian and British officials in the Curtiss proposal. At last, in April 1915, the British Admiralty requested that part of a contract placed with the Curtiss company for JN-3 training aircraft be completed in Canada. Curtiss agreed with this proposal and a new company was organized, Curtiss Aeroplanes and Motors Limited, Toronto, with McCurdy and Glenn Curtiss, another former AEA member and head of Curtiss Aeroplanes, as principal stockholders. A manufacturing plant was established in rented premises in Toronto, the first Canadian aircraft factory.

At around the same time, the Admiralty placed an order for twelve aircraft adapted as a land version of the large Curtiss flying boat H-1 *America*. This aircraft was named the Curtiss Canada and the order was fulfilled by Curtiss's Canadian subsidiary. Construction and test flying in 1915 went forward amid great secrecy and censorship, and all

Curtiss Aeroplanes & Motors Ltd. established Canada's first aircraft factory. During 1915 and 1916, it built eighteen Curtiss JN-3 trainers for the British Admiralty and for use at the company's flying school at Long Branch, Ontario.

twelve machines were sent to Britain in 1916. However, the design proved to be underpowered and the type had no clearly defined role in the evolving air war. Although a failure, the Curtiss Canada has the distinction of being the first twin-engine aircraft to be built in the country.

McCurdy continued campaigning diligently throughout 1916 to obtain additional contracts and to convince authorities to establish a Canadian Air Corps with training facilities in Canada. In spite of these efforts, the Canadian government—its financial resources strained by the war effort and its view of the post-war value of airplanes still skeptical—remained uninterested.

The concept of pilot training in Canada supported by Canadian-manufactured aircraft had been under consideration in Britain for many months when, late in December

ROYAL CANADIAN AIR FORCE
THE WRIGHT WHIRLWINDS
Now a CANADIAN Product
Avro NEWS
ARROW ROLLS OUT TO-DAY
IROQUOIS
ORENDA
ENGINES LIMITED · CANADA
ORENDA
AVRO CANADA
NOORDUYN AVIATION
N. H. BELL
FACTORY MANAGER
LIMITED
LINCOLN
VICTORY AIRCRAFT
WRIGHT
PRATT & WHITNEY AIRCRAFT
DEPENDABLE ENGINES

1916, a firm decision was taken to establish just such a program for the Royal Flying Corps, controlled and financed by British authorities. To support the program it was decided to manufacture training aircraft in Canada under the auspices of the Imperial Munitions Board. The board elected to establish a new company, Canadian Aeroplanes Limited, headed by F.W. (later Sir Frank) Baillie. The Curtiss manufacturing facilities were acquired for Canadian Aeroplanes Limited by expropriation, and the majority of Curtiss engineering and manufacturing personnel were transferred to the new company.

Canadian Aeroplanes immediately began developing an improved version of the Curtiss JN-3 for the Royal Flying Corps training program. Known as the Canadian JN-4 (Can.) or JN-4 Canuck, it was the first aircraft mass-produced in Canada and was supplied in large numbers to the Royal Flying Corps (RFC) program as well as to the U.S. Army Signal Corps. More than twelve hundred aircraft were completed during 1917–18, plus the spare-parts equivalent of an additional sixteen hundred machines, a remarkable record by any standard.

As Canadian Aeroplanes quickly outgrew the facilities taken over from Curtiss, a new factory was built in Toronto; transfer to the new plant was completed during May 1917. The U.S. government was favourably impressed with the record of this Canadian company, and a contract was signed with the U.S. Navy in early 1918 for the manufacture of fifty Felixstowe F-5-L flying boats, a large American version of a British-developed twin-engine aircraft. By 1919, thirty flying boats had been delivered to the U.S. Navy, the contract having been cut back at the war's end.

By 1918, less than two years after its establishment, Canadian Aeroplanes had developed into a large complex organization generally regarded as the most efficient aircraft manufacturer in North America. Many obstacles to production were overcome: sources of raw materials were developed, subcontractors located, new production techniques pioneered, and engineering and manufacturing personnel were trained. Then, almost overnight, all these hard-won skills and facilities were set aside. Canadian Aeroplanes and its

facing page: Artifacts from Canada's manufacturing past include a Vickers Vedette wind tunnel model of the early 1920s (upper right), a presentation model of the Avro Arrow given to the Minister of National Defence in the late 1950s (centre), and Canadian Aeroplanes Limited blueprints produced during the First World War.

above: A ceremony marking the delivery of six Curtiss-Reid Ramblers to the RCAF at Camp Borden, Ontario, in 1930. Second from left is J.A.D. McCurdy, company manager at the time.

above: Model C-204 Thunderbird flying boats, primarily designed as a fishery patrol aircraft, under construction at Boeing Aircraft of Canada Ltd., Vancouver, in 1930. A batch of four were built but did not sell, due to the onset of the Great Depression in 1929.

facing page: Canadian Vickers Vedette Mk Va over Montreal, by Robert Bradford. This Vedette is on a test flight over the Canadian Vickers factory in Montréal, where it was built. This company was the country's first post-war aircraft manufacturer, and versions of this small flying boat were first produced in the early 1920s for the RCAF.

British parent, the Imperial Munitions Board, were strictly wartime organizations. Few in the Canadian government or private sector recognized the development potential inherent in the aircraft industry.

Airplane use in Canada was, however, gradually increasing. The first civil forestry patrols were flown in 1919 from Grand-Mère, Québec, using Curtiss HS-2L flying boats; in 1920, a small Canadian Air Force was formed. Soon civil and military mail transport, forest-fire patrols, aerial exploration, and photo-survey operations commenced using war-surplus aircraft.

The industry revived when the Canadian Air Force ordered its first new aircraft, eight Vickers Viking amphibians. This encouraged the establishment of Canada's first post-war aircraft manufacturer, Canadian Vickers Limited. Its first two aircraft were built from parts supplied by the British parent firm; a further six were manufactured in Montréal and delivered to the Royal Canadian Air Force (RCAF) during 1923. The company then undertook designs that reflected Canadian requirements. The first was a small flying boat, the British-designed Vickers Vedette, whose production became the largest program of the interwar years.

Although RCAF and civilian funds available for aircraft purchase were limited during the 1920s, Canadian Vickers designed and constructed their own original types: the Varuna fire-suppression flying boat; the Vanessa light transport biplane; the Vista light forest-patrol flying boat; the Vigil forest-patrol biplane; the Velos photo-survey aircraft, and the Vancouver fire-suppression flying boat. Only two of these, the Varuna and the Vancouver, progressed past the prototype stage to limited production. In addition to their own designs, Canadian Vickers produced several types under licence. The Avro 504 and 552, the Fairchild FC-2, and the Fokker Super Universal were all produced in small numbers. Lack of success with their own designs and failure to obtain exclusive rights to supply RCAF aircraft caused Canadian Vickers to abandon original design work, and with the onset of the Depression in 1929 aircraft orders gradually dried up. An RCAF order in 1931 for six Bellanca Pacemakers was the last aircraft order they were to receive for five years.

VICKERS

Norseman construction at the Noorduyn Aviation Ltd. factory, Cartierville, P.Q., in 1941. Beginning in the mid-1930s, the Norseman was the first purpose-built bush plane to be designed and built in Canada.

In 1928, Vickers's chief engineer, W.T. Reid, left the company and set up his own operation, Reid Aircraft Company, at Cartierville, Québec. Reid designed and built a light airplane of tubular steel structure, the Rambler, an attractive and successful design based on preliminary work carried out at Canadian Vickers. The Curtiss Aeroplanes and Motor Company purchased controlling interest in the concern early in 1929, changing the name to Curtiss-Reid Aircraft Company; Rambler production then commenced. Unfortunately, the onset of the Depression later that year had a crippling effect on company prospects and only a small number of aircraft were completed. Curtiss-Reid also produced an attractive light mail plane, the Courier, but only a single prototype was built.

In 1929, the year of the Wall Street crash, Fairchild Aircraft Limited was established at Longueuil, Québec, near Montréal. A subsidiary of Fairchild Aviation, an American organization, it was to become the most successful aircraft manufacturer in Canada during the 1930s, producing bush planes for a growing bush-flying community. The first aircraft type produced was the Fairchild Model 71B. Then, in spite of the dismal economic outlook prevailing in the country, they designed an improved version, the Model 71C, and a new type, the Super 71. This new aircraft featured the first all-metal, stressed-skin fuselage to be built in Canada, but it was not a commercial success. However, Fairchild Canada went on to design and manufacture the Model 82, a successful bush aircraft well liked because of its load-carrying capabilities.

It is a measure of the precarious state of the industry in Canada during the 1930s that we can point to Fairchild as that decade's most successful manufacturer. Although they produced a total of only fifty aircraft from 1929 to 1939, they did nurture a number of experienced design and production personnel. By the outbreak of the Second World War, almost every Canadian aircraft manufacturer relied to some extent on these personnel.

Noorduyn workers in 1941 securing the fabric covering to a wing's underlying structure, a task known as rib-stitching. Later, about eight coats of a cellulose-based dope were applied to the fabric, making it taut and waterproof.

A number of other concerns managed to start up and struggle along during this period. On the west coast, Boeing Airplane Company set up a Vancouver subsidiary in 1930, Boeing Aircraft of Canada, which produced Canadian versions of their Models 204 flying boat and 40-B mail planes. Boeing Aircraft of Canada also produced one original design, the Totem flying boat, but by 1932 their manufacturing operation ceased for some years because of the Depression.

More successful was Fleet Aircraft Limited at Fort Erie, Ontario. This company was set up as a subsidiary of the Consolidated Aircraft Corporation, of Buffalo, New York, to manufacture their well-known Fleet trainers and to circumvent U.S. export regulations. They did a thriving business by the standards of the day, with sales to at least a dozen countries. Fleet eventually became a completely Canadian company when Consolidated relocated to San Diego, California. In addition to the Fleet trainer series, they built a few Model 21 trainers for export to Mexico and designed an original twin-engine utility transport, the Model 50 Freighter, five of which were produced.

above: Robert Bradford's painting *DH 60X Moth* depicts the flight that launched de Havilland Canada in 1928. Leigh Capreol test flew the first de Havilland aircraft to be assembled in Canada over the company's first assembly and storage building, a former canning shed.

facing page: Canadian Car & Foundry's chief engineer "Elsie" MacGill and plant manager John Soulsby pose with the company's Maple Leaf Mk. II trainer. The first aircraft designed by a female engineer, it received its certificate of airworthiness eight months after design work started.

It took great courage and optimism for anyone to start a new company during the 1930s, but an important new company appeared in 1934. Dutch-born Robert Noorduyn, with extensive experience in Britain and the United States, came to Canada and established a company of his own in Montréal to design and manufacture a utility transport, the famous Noorduyn Norseman. The Norseman, although very successful and destined to become a classic bush aircraft, sold only in small numbers until large RCAF and U.S. orders were received during the Second World War.

The British de Havilland Aircraft Company set up a small operation in Toronto during 1928 for overhaul and repair work, and it soon began assembly of aircraft using components from the parent company. By 1930 the Canadian factory was producing some minor components and started assigning de Havilland Aircraft of Canada construction numbers to Canadian-assembled DH.60 Moth trainers and DH.80 Puss Moth utility airplanes. In 1937, de Havilland Aircraft of Canada, destined to become one of the country's most successful aircraft companies, commenced complete manufacture of Tiger Moth trainers for the RCAF.

Perceiving the war clouds gathering in Europe, Canadian Car and Foundry Company Limited, an established Canadian railway equipment maker, decided in 1936 to establish an aircraft manufacturing facility at Fort William (now Thunder Bay), Ontario. They planned an ambitious program to manufacture other company's designs as well as to design and build original types. Their first original type was the Maple Leaf II, a biplane designed by Elsie MacGill, which flew in 1939 and was the first airplane designed by a woman engineer. Their first production work was assembly of fifty-two licence-built Grumman G-23 fighters for export, using American-manufactured components.

In 1939, Canadian Car and Foundry also produced an original biplane fighter, the FDB-1, better known as the Gregor after its designer Michael Gregor, the company's chief engineer. It was very attractive but was quite obviously obsolete when built. However, it served to show that the company had ability and potential.

Elizabeth Muriel Gregory MacGill (1905–1980)

"ELSIE" MACGILL was educated in Vancouver, at the University of Toronto, and at Michigan State University, where she graduated in 1929 with a master's in Aeronautical Engineering, becoming the first woman in the world to enter this field. At this time she was stricken with acute infectious myelitis, a form of polio. She wrote her final exams from a hospital bed, determined that her disability would not stop her pursuit of a career in engineering. After post-graduate work at the Massachusetts Institute of Technology in the early 1930s, she was hired as an engineer at Fairchild Aircraft Limited, Longueuil, Québec, where she worked on stress analysis of the prototype Fairchild Super 71, the first stressed-skin, all-metal aircraft designed and built in Canada.

In 1938 she left Fairchild to become chief aeronautical engineer at the Fort William plant of Canadian Car and Foundry Limited. There, she became the first female aeronautical engineer to work on the overall design of an airplane, the Maple Leaf II, which most represented her originality. A two-seat, single-engine primary trainer designed for use on wheels, skis, or floats, it received its certificate of airworthiness within eight months of the commencement of design, an outstanding achievement. Unhappily, it proved to be so gentle to fly that it was insufficiently challenging to potential pupils. It would have made a pleasant leisure machine in peacetime, but it was not what was needed in 1939–40.

At the beginning of the Second World War, Canadian Car and Foundry began large-scale production of military aircraft. MacGill was put in charge of all engineering work related to production of the British-designed Hawker Hurricane fighter, and within a year, three aircraft a day were being produced for service overseas. Her wartime work also involved winterizing the Hurricane for Canadian conditions, and doing production-line design work for the 835 Curtiss Helldivers that the firm built under contract to the U.S. Navy.

In 1946 she became the first woman to serve as Canadian technical adviser to the United Nations International Civil Aviation Organization (ICAO), where she helped draft international airworthiness regulations for the design and production of commercial aircraft. Elsie MacGill's accomplishments earned her wide recognition and many honours, including honorary doctorates from four Canadian universities. Subsequently she served on the Royal Commission on the Status of Women, and she was appointed an Officer in the Order of Canada in 1971.

above: Canada's aircraft industry employed approximately 3,600 people in 1939, and total aircraft orders numbered less than 200. By 1945, employment had surpassed 120,000, and 16,448 aircraft had been built. The English version of this poster reads "Roll 'em out."

right: The manufacture of dated designs such as these Bristol Bolingbrokes provided the know-how that would enable Canada's aircraft industry to begin mass production of advanced designs like the de Havilland Mosquito and Avro Lancaster.

During the late 1930s, a few small companies were in existence, all with limited capital resources, only a small pool of trained, experienced engineering and production personnel, and virtually no network of subcontractors or raw material sources. The companies were making efforts to interest the British and Canadian governments, but with little success. As the international situation darkened, the Canadian government finally began to consider re-equipping the RCAF on a small scale.

Contemporary posters reinforced the important role played by the factory worker during the Second World War in the march towards ultimate victory.

Northrop Delta light transports and Supermarine Stranraer flying boats were ordered from Canadian Vickers, Blackburn Shark torpedo-bombers from Boeing Aircraft of Canada, Fawn trainers from Fleet Aircraft, Tiger Moth trainers from de Havilland Aircraft of Canada, and Bristol Bolingbroke coastal patrol airplanes from Fairchild Aircraft. In 1938, National Steel Car Corporation formed an aircraft division at Malton, Ontario, to produce Westland Lysander observation aircraft for the RCAF. None of these orders was particularly large but even so, they severely taxed the limited available industrial capacity. All the combat types ordered proved obsolete when delivered and saw limited service in their intended operational roles. However, they proved useful as trainers during the war and, more importantly, they allowed the various companies to begin increasing production capacity and staff for the war effort.

On the eve of the Second World War, the British government overcame its amused tolerance of Canadian proposals to manufacture aircraft for the Royal Air Force (RAF). In 1938, they placed an order for Handley Page Hampden bombers with Canadian Associated Aircraft Limited, a consortium of six manufacturers who pooled their resources to provide the financial stability required of the program. A separate RAF contract included the production of Tiger Moth fuselage structures, many of which were delivered to the RCAF in 1940–41.

In September 1939, an agreement was reached between British and Commonwealth governments to set up a large-scale training program in Canada, the British Common-

facing page: An Avro Anson Mk II fuselage being married to its one-piece wing at the National Steel Car plant in Toronto near the beginning of the Second World War. Some of the problems initially encountered in producing the British-designed Anson were caused by the use of both North American and British hardware standards in its construction.

wealth Air Training Plan (BCATP—see previous chapter). To support this program, large orders were placed for training aircraft: Tiger Moths from de Havilland Canada, Finch trainers and subsequently Fairchild Cornells from Fleet, and Norseman and North American Harvards from Noorduyn. Early production difficulties are exemplified by the production of the Avro Anson, a British design selected as the standard twin-engine trainer for the BCATP. A Crown company, Federal Aircraft Limited, was formed to co-ordinate a group of aviation and non-aviation companies in Anson production. Sources for engines, propellers, instruments, hardware, raw materials, and equipment of all types had to be located or developed. Some items, for example, had to be procured in the United States. To add to the difficulties, engineering changes were required to adapt the British aircraft to North American standards. With eleven major contractors and almost two hundred subcontractors, all seemed chaos for a time. Eventually, the Anson V was produced, and what emerged was virtually a new, Canadian-designed airplane.

By 1943, aircraft production was running relatively smoothly and remarkable records were being set. Production of Ansons, Harvards, and Cornells was progressing well. Canadian Car and Foundry and Fairchild Aircraft were between them turning out the first of 1,135 Curtiss Helldivers for the U.S. Navy, while de Havilland production of Mosquitos was gaining momentum. Boeing and Canadian Vickers were producing Consolidated PBY-5 Catalina flying boats and PBY-5A Canso amphibians, used by the RCAF as well as other allied forces. In Toronto, National Steel Car's aircraft division had become a Crown corporation, Victory Aircraft Limited, and was well along with another large Canadian program, the production of 430 Avro Lancaster bombers. Under the supervision of Elsie MacGill, Canadian Car and Foundry also produced 1,450 Hawker Hurricane fighters.

During 1943–44, a development occurred that was unique in Canadian industrial history. C.D. Howe, the powerful and controversial but nonetheless great cabinet minister, began to consider the post-war future of the aviation industry in Canada. It was quite apparent that at the war's end the industry would be drastically reduced. The government decided that this time it would intervene to prevent its complete disappearance, as had happened with Canadian Aeroplanes Limited in 1918. As Victory Aircraft in Toronto was already a Crown company and Canadian Vickers in Montréal was operating in a new government-built factory, these two companies were chosen for protection. At this time,

Trans-Canada Air Lines and the Canadian government decided to develop a Rolls-Royce Merlin–powered version of the Douglas DC-4 airliner. The government reorganized Canadian Vickers' Cartierville factory as a Crown company, Canadair Limited, to produce the new transport for the RCAF and Trans-Canada Air Lines (TCA).

By the end of the Second World War, the industry could look with pride on its substantial achievements. A large number of secondary manufacturers were producing components such as instruments, propellers, and hydraulic and electrical parts for which there had been no Canadian source in 1939. There were still significant gaps in capacity: no aircraft engines were being manufactured, and of all the types of aircraft manufactured during the war only two, the Noorduyn Norseman and Fleet Fort trainer, were of Canadian design. Although aircraft production was Canada's fourth-largest industry by the end of the war, engineering personnel, particularly designers, were still few and far between. Both these deficiencies were made good in the immediate post-war years.

CL-215T
239
Québec
C-FAWO

facing page: Critical Approach, by Charles Vinh. Uniquely Canadian post-war success stories, the Canadair CL-215 and CL-415 have become well known throughout the world. These water bombers can repeatedly disgorge tons of fire retardant on a burning forest by scooping water from nearby lakes or rivers in one pass.

above: The Canadair CP-107 Argus, the largest aircraft in the Museum's collection, served as an anti-submarine and maritime reconnaissance aircraft. Powered by four Wright turbo-compound engines of 3,700 horsepower each, it had a range of 7200 kilometres (4,500 miles) and an endurance of 26 hours.

POST-WAR CANADIAN PRODUCTION: CANADAIR, AVRO, AND DE HAVILLAND

One lesson learned during the war was that Canada's military could not necessarily get the aircraft it wanted when other parties were competing for the same machines. Thus, although the aircraft industry underwent drastic reduction and reorganization after the war, the Canadian government itself was sympathetic to Canadian designs and production for specific domestic needs. Then, as world tension between East and West increased in the late 1940s, particularly with the outbreak of the Korean War in 1950, re-equipment orders for the RCAF again flowed in, making the 1950s a happy and busy period for the industry.

The first major step in converting to peacetime operations involved the transformation of Canadian Vickers aircraft division into government-owned Canadair in 1944. It completed the last of wartime orders of the original company and then set about adapting the Douglas DC-4 to the Rolls-Royce Merlin engine, a marriage resulting in the North Star, which first flew in July 1946. The company went on to adapt the North American Sabre to a Canadian-designed jet engine (the Orenda), creating a line of fighter aircraft that were the envy of North Atlantic Treaty Organization (NATO) air forces for a decade. Again following a Canadair tradition, the company adapted a British airliner (the Bristol Britannia) to both an anti-submarine role (the CP-107 Argus, test flown in March 1957) and a transport role (the CL-44, November 1959). It also adapted an American airframe (the Convair CV-440) to a British turboprop engine (the Napier Eland) to create the Canadair CL-66 Cosmopolitan (January 1960). In addition, the company manufactured Lockheed T-33 jet trainers, Lockheed CF-104 Starfighters, and Northrop CF-5 fighters for RCAF and NATO use.

The Argus is an example of how extensive the modifications needed to adapt an aircraft to a new role could be. The turboprop engines of the Bristol Britannia on which it was based were relatively inefficient for low-level cruising, but the Argus had to operate close to the sea to find, identify, and if necessary attack targets. The original Bristol Proteus power plants were replaced by Wright R-3350 turbo-compound piston engines. Fitting a radar dome in the nose meant redesign of the landing gear, and incorporating magnetic anomaly detection (MAD) gear in the tail meant revision of the rear fuselage. The vertical fin was modified so that it could also function as a radio antenna, and the entire airframe was strengthened with titanium and high-strength aluminum alloys.

The Avro Canada C-102 Jetliner was the first jet transport designed and built in North America. Despite its promise, requirements for CF-100 fighter production stopped its development, and only its cockpit section has survived, housed at the Canada Aviation Museum. Designer Jim Floyd is second from right.

Adaptation was one form of development, but Canadair also struck its own line of uniquely Canadian aircraft. The company designed and manufactured the CT-114 Tutor jet trainer (1960), the CL-84 Dynavert tilt-wing vertical-takeoff-and-landing (VTOL) aircraft (1965), and the CL-89 unmanned reconnaissance drone. Canadair's world-famous CL-215 water bomber (1967) and its turboprop adaptation, the CL-415 Super Scooper (1993), have become familiar sights in Europe and North America as televised news coverage of dramatic forest fires routinely shows them swooping down to disgorge tons of water and chemical retardants. The CL-600 Challenger executive jet transport (1978) represented a return to Canadian development and manufacture of a foreign design, although the Challenger was very different from the original American design.

After a period of American ownership (1947 to 1976), the company was again acquired by the Canadian government and then sold back to the private sector in 1986. Today it is a division of Bombardier Aerospace.

In 1945, Avro Aircraft Limited was formed in Canada as part of the Hawker Siddeley Group (U.K.) through the purchase of the government-owned Victory Aircraft. Unlike Canadair, the company embarked upon a line of original designs. Their first project, the C-102 Jetliner, was test flown in 1949 but failed to receive commercial orders, although it had the distinction of being the world's second jet airliner, behind Britain's de Havilland Comet. Due to the uncertainties of the early Cold War years, in 1950 the Canadian government pressured Avro to concentrate on the production of the CF-100 straight-winged fighter instead. Avro also entered the power-plant field, designing and manufacturing the Orenda series of jet engines.

In 1951, Avro and Orenda began development of a new fighter and engine to replace the CF-100. The aircraft that resulted was the Avro CF-105 Arrow, the first and last supersonic interceptor designed and built in Canada. The RCAF specification requirements issued for the project in 1953 were staggering and surpassed anything documented in any other country in the Western world. They called for an interceptor that could

The Beaver Prototype

PRIOR TO designing the Beaver in 1946, de Havilland Canada canvassed potential operators, particularly the Ontario Provincial Air Service, for desired specifications of the ideal bush plane. First flown in August 1947, the Beaver was a success, although it had to compete with war-surplus Noorduyn Norsemans that lacked the Beaver's short-takeoff-and-landing (STOL) qualities. Ultimately, 1,632 Beavers and 60 Turbo-Beavers were built, the Beaver becoming the world's best-known bush plane.

In 1980, the Canada Aviation Museum was fortunate to acquire the very first Beaver from its last operator, Norcanair of Prince Albert, Saskatchewan.

The DHC-2 Beaver prototype taxiing on the Ottawa River after its delivery flight to the Museum in 1980. The letters of its registration "FHB" were allocated to honour one of its designers, Fred H. Buller.

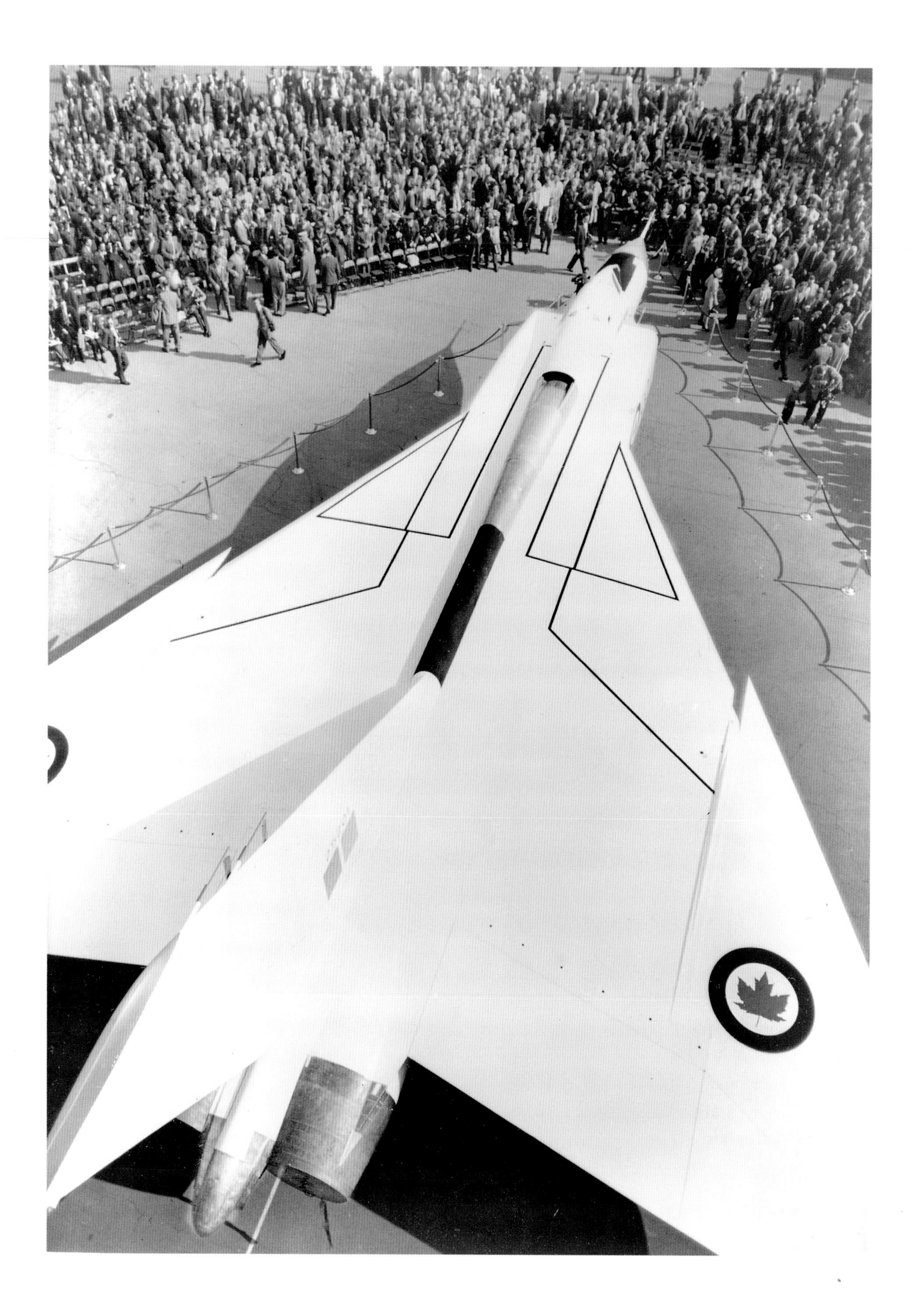

facing page: The Avro CF-105 Arrow on the day of its roll-out ceremony, October 4, 1957. Ironically, this was also the day the Soviet Union launched its Sputnik satellite, alerting the Western world to its capabilities with rocket technology. The anticipated production of intercontinental ballistic missiles became a factor in the Arrow's eventual cancellation.

operate from a 1830-metre (6,000-foot) runway, have a range of 1120 kilometres (690 miles), and be capable of accelerating to Mach 1.5 with a crew of two, carrying a very advanced type of missile with a control system capable of operating in a harsh environment. The missile system would require a weapons bay under the aircraft larger than those found on many Second World War bombers.

The project proceeded quickly, with the company skipping the prototype stage and going straight to production tooling; the first flights were conducted by early 1958. Designed to counter the perceived threat of supersonic Soviet bombers entering Canadian airspace over the polar regions, the aircraft's advanced performance became a symbol of national pride. However, air forces globally were abandoning the concept of single-purpose interceptors in favour of multi-role aircraft with sophisticated air-to-air and air-to-ground weapons.

When it became obvious that intercontinental ballistic missiles were making nuclear bombers obsolete—and hence also aircraft designed to intercept bombers—the Arrow's redundancy was recognized. Its Iroquois engine and weapons system were also extremely expensive (even its original sponsors had misgivings about rising development and projected manufacturing costs) and a new Canadian government, elected in 1957 with no ties to the relatively cohesive industrial policy followed by its predecessors, cancelled the entire project on February 20, 1959, a day known in the industry as Black Friday.

This action effectively put Avro out of business and its highly skilled engineering and production personnel scattered, some to other Canadian companies but most to the United States. The incident was a traumatic one, as many subcontractors had also been fatally hit. At the time, many people predicted the end of the industry in Canada, and to this day many mourn the loss of the Arrow.

De Havilland Canada emerged from the war with a plan to produce machines of their own design. Displaying remarkably good business sense, the company commenced the post-war era by producing Fox Moth biplanes, which became popular with bush operators. This 1932 British type was manufactured using many surplus Tiger Moth components and allowed the company time to produce their first original designs, the DHC-1 Chipmunk trainer and the DHC-2 Beaver utility transport. These were very successful, as was the larger DHC-3 Otter that followed. From the 1950s onward, de Havilland *facing page:* The prototype Grumman CS2F Tracker, the first anti-submarine aircraft to be built in Canada for the Royal Canadian Navy, passes over the plant where it was assembled in 1956. De Havilland Canada was the prime Canadian contractor for this American design, but many of its components were subcontracted to other Canadian companies.

continued to develop a successful line of short-takeoff-and-landing (STOL) transports with the DHC-4 Caribou, DHC-5 Buffalo, DHC-6 Twin Otter, and Dash 7 STOL airliner, followed by the Dash 8 regional airliner. These well-known designs have sold in substantial numbers to civil and military operators around the world and made de Havilland Canada one of the country's most successful aircraft manufacturers.

Several pre-war companies were less successful in the post-war period. Noorduyn commenced production of a modernized Norseman, but they were competing with large numbers of surplus aircraft of the original model and soon encountered financial problems. They sold Norseman manufacturing rights to Canadian Car and Foundry and went out of business in 1946.

Fairchild Aircraft had built a promising bush plane before the Second World War (the Fairchild 82), which might have rivalled the Noorduyn Norseman. The company concentrated on building Bristol Bolingbrokes during the war and apparently scrapped the tools needed to resume Fairchild 82 production afterwards. In 1945 the company undertook design of a new aircraft, the F.11 Husky, to carry on their pre-war tradition of building bush aircraft. However, the Ontario Provincial Air Service compared the Husky unfavourably with the Beaver, bought the latter, and thus doomed Fairchild Aircraft itself.

Fleet Aircraft took longer to fade away as a complete aircraft manufacturer. The company had high hopes for a post-war light aircraft, the Fleet 80 Canuck, for which it had purchased the manufacturing rights from its designer, Robert Noury. Produced as the Fleet Canuck, two hundred were sold, mostly in Canada. However, Fleet had overestimated the post-war market and underestimated the competition offered by Aeronca, Cessna, and Piper aircraft. It sold manufacturing rights to Leavens Brothers of Toronto (which built twenty-six more Canucks). By 1973, Fleet was building components assembled elsewhere for other firms and later became a component of Magellan Aerospace.

Canadian Car and Foundry turned out one odd-looking transport, the CBY-3 Loadmaster, but was not successful in marketing it. After securing the rights to the Noorduyn Norseman, the company assembled thirty-three of them between 1946 and 1953 and then made an attempt to modernize the basic design with an all-metal wing and other improvements, producing the Norseman VII. Unfortunately, the project was abandoned when a large Harvard trainer order was received from the RCAF in 1950. The company subsequently produced Beechcraft T-34 Mentor trainers for the RCAF and for export. Its aircraft division was eventually purchased by Avro Canada in 1955.

AIRCRAFT ENGINE PRODUCTION IN CANADA

From the 1920s to the end of the Second World War, the Canadian aircraft industry was handicapped by the absence of a corresponding aero-engine industry. Subsidiary companies such as Canadian Pratt & Whitney Aircraft and Canadian Wright established repair and maintenance facilities but did not assemble power plants. Consequently, Canadian-built airframes were dependent on engines manufactured in Britain or the United States. This was not a problem in peacetime, but it complicated wartime deliveries when engine supplies from Britain were uncertain and there was intense competition for American engines.

As a new form of power appeared—the jet engine—Canadians took steps to ensure the nation would be involved in the new technology. Winnett Boyd, a scientist with the National Research Council, was in the forefront of these moves. In 1943 he was dispatched to Britain to study gas turbine engines with Sir Frank Whittle, Britain's jet-engine pioneer, as his mentor. Upon his return to Canada, Boyd joined a short-lived Crown corporation, Turbo Research Limited, as chief designer. Following Avro's acquisition of the firm, he developed the Chinook engine as a test bed, but it was never actually flown. It was, however, the first step on the way to the famous Orenda, and it ushered Canada into the jet age.

While Orenda Engines led the way in Canadian engine design and manufacturing, today's Pratt & Whitney Canada (PWC), a United Technologies company, remained essentially a servicing branch of its American parent until 1952, when it became an engine manufacturer in its own right, Canadian Pratt & Whitney Aircraft, building R-1340 Wasp radial engines for Harvards manufactured at Fort William, Ontario, for the RCAF. Although production of these engines ceased in 1956, the company carried

above: Company officers of Avro's Gas Turbine Division viewing the Canadian-designed Orenda Chinook engine on its test bed, with a visiting Sir Frank Whittle second from left. The Chinook first ran in 1948.

facing page: The Iroquois, planned successor to the Orenda, with an anticipated thrust of 13 150 kilograms (29,000 pounds) with afterburner, was the most powerful engine produced in North America at the time. The sixth and subsequent Avro Arrows were to be powered by this engine.

on building under licence Wright R-1820 Cyclones that were destined for Grumman CS2F Trackers ordered by the Royal Canadian Navy. The production of these engines gave PWC further experience building engines and more time to establish a commercial business.

Although the era of the piston engine was drawing to a close, the overhaul of radial engines and the profitable business of parts supply would last a few more years. Preparing for the new age, the company embarked upon the design of a lightweight gas turbine that evolved into the PT-6A in 1959. This radical new design changed the face of aviation. Powering the DHC-6 Twin Otter and Dash 7, derivatives of this small turbine have found applications in a wide range of commuter and regional aircraft worldwide.

Building on the PT-6A's success, the company diversified and expanded its line of engines, with new models including the JT-15D turbofan, a pioneer design in the category of small fan jets (which power a number of executive business jets), the PW-100 turboprop, the PW-200 turboshaft for helicopters, and the PW-300, -500, and -600 turbofans. Today, Pratt & Whitney Canada is the world's largest manufacturer of small-to-mid-range turbine engines.

Beginning in 1975, a new chapter opened in Canada's aerospace industry. Due to rather reluctant backing—financial and otherwise—from their British and American owners, de Havilland Canada and Canadair faced an uncertain future, with the Dash 7 airliner and CL-600 Challenger executive jet projects in doubt. To ensure their future, the Canadian government purchased the two companies, with the stated intention of returning them to private ownership as soon as possible.

Following the federal elections of 1984, the government announced its intention to sell de Havilland Canada to the private sector, and the company was eventually purchased by Boeing in 1986. The Boeing years produced significant benefits for engineering at de Havilland Canada's facility as costly upgrades to the plant were carried out, including computerized numerical control equipment, automatic riveting machines, increased capacity for making composites, and computerized manufacturing. The stretched version of the Dash 8 (Dash 8-300) was accomplished under Boeing's banner. Unfortunately, the Boeing/de Havilland union eventually led to disillusionment.

Orenda Engines

THE THREE Chinook engines built under the direction of Winnett Boyd developed some 1360 kilograms (3,000 pounds) of thrust and were a learning experience for Canadian aero engine designers. The Gas Turbine Division of Avro Canada (later Orenda Engines) set about designing an engine for future RCAF fighters, and in 1949 tests commenced on what would become the Orenda, named for an Iroquoian deity. The prototype CF-100, flown in January 1950, had Rolls-Royce Avon engines, but subsequent models used the Orenda, as did Canadair Sabres from 1953 onward. Some four thousand Orendas were built, with production ceasing in 1958. Typical of these engines was the Orenda 11, which delivered 3310 kilograms (7,300 pounds) of thrust.

The planned successor to the Orenda, the Iroquois, was projected to produce an estimated 13 150 kilograms (29,000 pounds) of thrust, making it one of the most powerful engines produced in North America to that date. It was cancelled with the Avro Arrow in 1959, but Orenda Engines survived the demise of Avro Canada, building American engines under licence for Canadair aircraft. The company later became a subsidiary of Magellan Aerospace. The Museum holds one of the three surviving Iroquois engines.

Laurent Beaudoin (1938–)

THE NAME Laurent Beaudoin is synonomous with the Québec company Bombardier Inc. This Canadian industrialist has changed the face of business in this country. Beaudoin, who is a Companion of the Order of Canada, transformed the family business, Bombardier Inc., into a multi-billion-dollar international operation with a global presence and interest in aerospace, motorized recreational vehicles, passenger rail cars, and transportation systems as well as financial services.

The son-in-law of company founder and snowmobile inventor J. Armand Bombardier, Beaudoin joined the company as comptroller in 1963 and took over soon thereafter. Employing diversification strategies throughout the 1970s, and continuing to take advantage of the Canadian aerospace industry's consolidation in the 1980s and early 1990s, he directed Bombardier's spectacular growth. Under his leadership, the company took many risks, including the acquisition of troubled aerospace companies such as Canadair, de Havilland Canada, and Short Brothers in Northern Ireland. After turning these companies around, Beaudoin asserted that "where others saw problems, we saw opportunities." Bombardier's first big decision as an aerospace entity, in 1989, was to develop and launch the fifty-passenger Canadair Regional Jet. That bet—a gutsy move by anyone's standards—wagered $250 million on a product for which many industry observers proclaimed there was no market.

With the acquisition of the American executive jet maker Learjet, Bombardier business groups, whose traditions dated from the very earliest days of aircraft manufacturing, became linked resources capable of collaborating on the most complex aircraft programs. Under Beaudoin's leadership, Bombardier Aerospace became the world's third largest civil aircraft manufacturer through acquisition, innovation, new products, strategic partnerships, geographic diversification, and careful targeting of regional and business aircraft markets. Beaudoin emerged as one of Canada's most respected and influential corporate leaders in the aerospace industry.

In the 1990s, Bombardier launched stretched versions of the de Havilland Dash 8 regional airliner. The 37-seat Q200, 50-seat Q300, and 70-seat Q400 comprise Bombardier's Q (Quiet) series of aircraft, named to reflect their revolutionary noise- and vibration-suppression systems.

The furor surrounding the second sale of de Havilland in six years brought media cries for a truly Canadian solution. Bombardier of Montréal was named as the logical buyer. Eventually, Bombardier saved the company and became the buyer of last resort, obtaining generous federal and provincial government benefits in the process. Under Bombardier, the Q (Quiet) series, its stretched versions of the Dash 8, were launched, introducing a whole new generation of turboprops with their revolutionary noise and vibration suppression (NVS) system.

The enterprising transportation manufacturer also turned Canadair around after buying it in 1986. Having acquired the company's interests and projects, it became associated with the Canadair Regional Jet series, initially a stretched version of the business jet that has been popular with both airlines and corporations; more than a thousand had been sold worldwide by 2004. Bombardier also acquired Short Brothers PLC in Northern Ireland in 1989 and Learjet in the United States in 1990. In 2004 the company's products included three versions each of the Canadair Regional Jet (CRJ) and Q Series (Dash 8) aircraft and the Challenger and Global Express business jets. Bombardier also builds the turbine-powered Bombardier 415 amphibious water bomber, using engines built by Pratt & Whitney Canada.

The evolution of an industry is always complex and seldom steady. The Canadian aircraft industry has been no exception. Since the formation and brief life of the Canadian Aerodrome Company in 1909, the industry has experienced every possible situation, from times of total extinction to periods of depression to exhilarating years of all-out production. Nevertheless, at the end of the twentieth century, Bombardier was the world's third-largest manufacturer of civil aircraft, Pratt & Whitney Canada the largest builder of small turbine engines, and Canadian Aviation Electronics (CAE) the pre-eminent constructor of sophisticated flight simulators. Also, Bell Helicopter Textron Canada Limited was the world's leading manufacturer of civilian helicopters. In short, the Canadian industry had become a world leader in several important aviation fields. Canadians are justly proud of the industry's achievements and look forward with confidence to the future.

Appendix

AIRCRAFT IN THE CANADA AVIATION MUSEUM COLLECTION

THE FOLLOWING IS a complete alphabetical list of aircraft in the Canada Aviation Museum's collection. The name of each aircraft is followed by the name of the manufacturer and date of manufacture. Construction and registration numbers follow where known. Finally, a brief note is included on how and when the aircraft came to be in the collection. For further information on these aircraft, consult the Museum's website at www.aviation.technomuses.ca.

Advanced Aviation Buccaneer SX (kit), U.S., built by Buzzman Enterprises Inc., Canada, 1992. Serial number 9112SX-AA; registration C-IDWT (Can.). Acquired as a donation in 2003 (Mr. Arleigh Vincent, Ottawa, Ontario).

AEA *Silver Dart* (reproduction), RCAF, Canada, 1957. Transferred from the RCAF in 1964. Second of three reproductions built.

AEG G.IV, Allgemeine Elektrizitäts Gesellschaft, Germany, 1918. Construction number unknown; registration 574/18 (German Air Service). Transferred from the Canadian Armed Forces in 1970.

Aeronca C-2, Aeronautical Corporation of America, U.S., 1931. Construction number 9; registration N525V (U.S.), CF-AOR (Can.; fictitious identity). Purchased from a private owner in 1967.

Airspeed Consul, Airspeed Ltd., U.K., 1945. Construction number 4338; registration PK286 (RAF), G-AIKR (U.K.). Purchased by the Canadian War Museum in 1965 and later transferred to the Canada Aviation Museum. On long-term loan to the Royal New Zealand Air Force Museum.

Apco Astra 29 (paraglider), Apco Aviation Ltd., Israel, 1992. Serial number 300316. Acquired as a donation in 1999 (Mrs. Vincene Muller, Cochrane, Alberta).

Auster AOP 6, Auster Aircraft Ltd., U.K., 1948. Construction number 2576; serial number TAY-221V; registration 16652 (RCAF), CF-KBV (Can.), VF582 (RAF; fictitious identity). Purchased by the Canadian War Museum in 1965 and later transferred to the Canada Aviation Museum.

Avro 504K, A.V. Roe and Company Ltd., U.K., c. 1918. Construction number 2353; registration H2453 (RAF), 5918 (U.S.), N8736R (U.S.), G-CYFG (RCAF; fictitious identity). Transferred from the RCAF in 1967.

Avro 504K, Graham White Aviation Co., U.K., 1918. Construction number unknown; registration D8971 (RAF), G-CYCK (RCAF; fictitious identity), CF-CYC (Can.). Transferred from the RCAF in 1968.

Avro Anson V, MacDonald Brothers Aircraft Ltd., Canada, 1944. Construction number MDF329; registration 12518 (RCAF). Transferred from the RCAF in 1964.

Avro Avian IVM, Ottawa Car Manufacturing Co., Canada, 1930. Construction number R3/CN/314; registration 134 (RCAF), CF-CDQ (Can.). Acquired as a donation in 1968 (Mr. Charles Graffo, Winnipeg, Manitoba).

Avro Canada C-102 Jetliner (forward fuselage), A.V. Roe Canada Ltd., Canada, 1949. Construction number unknown; registration CF-EJD-X (Can.). Donated to the National Research Council in 1956 by Avro Aircraft and later transferred to the Canada Aviation Museum.

Avro Canada CF-100 Mk.5 Canuck, A.V. Roe Canada Ltd., Canada, 1958. Construction number 685; registration 100785 (CAF; fictitious colours and markings). Transferred from the Canadian Armed Forces in 1982.

Avro Canada CF-100 Mk.5 Canuck, A.V. Roe Canada Ltd., Canada, 1958. Construction number 657; registration 100757 (CAF). Transferred from the Canadian Armed Forces in 1979.

Avro Canada CF-105 Arrow Mk.2 (forward fuselage), A.V. Roe Canada Ltd., Canada, 1959. Construction number 206; registration 25206 (RCAF). Transferred from Defence and Civil Institute of Environmental Medicine in 1965.

Avro Lancaster X (forward fuselage), Victory Aircraft Ltd., Canada, c. 1944. Construction number unknown; registration KB 848 (RCAF). Transferred from the RCAF in 1966.

Avro Lancaster X, Victory Aircraft Ltd., Canada, 1945. Construction number unknown; registration KB 944 (RCAF; fictitious markings). Transferred from the RCAF in 1964.

Bell CH-135 Twin Huey, Bell Aircraft Corporation, U.S., 1971. Construction number unknown; registration 135114. Transferred from the Canadian Forces in 1998.

Bell HTL-6 (47G), Bell Aircraft Corporation, U.S., 1955. Construction number 1387; registration 1387 (RCN). Transferred from the Royal Canadian Navy in 1966.

Bellanca CH-300 Pacemaker, Bellanca Aircraft Corporation, U.S., 1929. Construction number 181; registration NC196N (U.S.), CF-ATN (Can.). Purchased in 1964.

Bensen B8 Gyroglider, Bensen Aircraft Corporation, U.S., c. 1977; kit assembled by Alexander Dutkewych, Toronto. Serial number unknown; registration C-GSXV (Can.). Acquired as a donation in 2002 (Mr. Alexander Dutkewych, Toronto, Ontario).

Bensen B8MG Gyrocopter, Bensen Aircraft Corporation, U.S., 1974. Serial number P-6174; registration C-GPJE (Can.). Acquired as a donation in 2002 (Mr. Alexander Dutkewych, Toronto, Ontario).

Blériot XI, California Aero Manufacturing and Supply, U.S., 1911. No construction number; no registration. Purchased in 1971.

Boeing 247D, Boeing Airplane Co., U.S., 1934. Construction number 1699; registration NC13318 (U.S.), CF-BAS (Can.), 7638 (RCAF), CF-BVX (Can.), NC41809 (U.S.), CF-JRQ (Can.). Acquired as a donation in 1967 from California Standard Oil Co., Calgary, Alberta.

Boeing MIM-10B Super Bomarc (missile), Boeing Airplane Co., U.S., 1960. Construction number 656; registration 60446 (RCAF). Transferred from the Canadian Armed Forces in 1972.

Boeing Vertol CH-113 Labrador, Boeing Airplane Co., Vertol Division, U.S., 1963. Construction number unknown; registration 11301 (CF). Transferred from the Canadian Forces in 2004.

Borel Morane, Société anonyme des aéroplanes Morel–Borel–Saulnier, France, 1911–12. Construction number unknown. Purchased in 2002.

Bowers Fly Baby 1A (incomplete), designed by Peter Bowers, U.S., 1960, homebuilt by Bill Loftus, Manotick, Ontario, Canada, 1970–2000. Acquired as a donation in 2002 (Mrs. Mary Loftus, Manotick, Ontario).

Bristol Beaufighter TFX, Bristol Aeroplane Co. Ltd., U.K., 1945. Construction number unknown; registration RD867 (RAF). Acquired in 1969 through exchange.

Bristol Bolingbroke IVT, Fairchild Aircraft Ltd., Canada, 1942. Construction number unknown; registration 9892 (RCAF), 9025 (RCAF; fictitious identity). Transferred from the RCAF in 1964.

Canadair C-54GM North Star I ST, Canadair Ltd., Canada, 1948. Construction number 122; registration 17515 (RCAF). Transferred from the RCAF in 1966.

Canadair CF-116A (CF-5A), Canadair Ltd., Canada, 1970. Construction number unknown; registration 116763 (CF). Transferred from the Canadian Forces in 1997.

Canadair CL-84-1 Dynavert, Canadair Ltd., Canada, 1969. Construction number 3; registration CX8402 (CAF). Acquired as a donation in 1984 from Canadair.

Canadair CP-107 Argus 2, Canadair Ltd., Canada, 1960. Construction number 33; registration 10742 (CAF). Transferred from the Canadian Armed Forces in 1982.

Canadair CT-114 Tutor, Canadair Ltd., Canada, 1963. Serial number 1108; registration 26108 (RCAF), 114108 (CF). Transferred from the Canadian Forces in 1999.

Canadair CT-133 Silver Star 3, Canadair Ltd., Canada, 1957. Construction number T-33-574; registration 21574 (RCAF). Transferred from the RCAF in 1964.

Canadair Sabre 6, Canadair Ltd., Canada, 1955. Construction number 1245; registration 23455 (RCAF). Transferred from the RCAF in 1964.

Canadair Sabre 6, Canadair Ltd., Canada, 1956. Construction number 1441; registration 23651 (RCAF). Transferred from the RCAF in 1964.

Canadian Vickers Vedette V (nose section only), Canadian Vickers Ltd., Canada, c. 1929. Construction number CV-146; registration G-CYVP (RCAF). Source of acquisition unknown.

Canadian Vickers Vedette VA (hull remains only), Canadian Vickers Ltd., Canada, c. 1929. Construction number CV-155; registration G-CYWQ, 814 (RCAF). Salvaged from Ontario crash site in 1977.

Cessna Crane, Cessna Aircraft Co. Inc., U.S., 1941. Construction number 2226; registration 8676 (RCAF). Transferred from the RCAF in 1964.

Consolidated B-24L Liberator GR VIII, Ford Motor Co., U.S., 1945. Construction number 44-50154 (USAAF); registration KN820 (RAF), 11130 (RCAF; fictitious identity). Acquired in 1968 through exchange.

Consolidated PBY-5A Canso A, Canadian Vickers Ltd., Canada, 1944. Construction number CV-423; registration 11087 (RCAF; fictitious colour scheme). Transferred from the RCAF in 1964.

Curtiss HS-2L, Curtiss Aeroplane and Motor Co., U.S., 1918. Construction number 2901-H.2; registration A1876 (USN), G-CAAC (Can.). Salvaged in 1968 from a northern Ontario lake by the Canada Aviation Museum. Aircraft reproduced incorporating parts of three aircraft, including new hull.

Curtiss JN-4 (Can.) Canuck, Canadian Aeroplanes Ltd., Canada, 1918. Construction number unknown; registration 39158 (U.S. Air Service), 111 (U.S. civil), C227 (RFC [Canada]; fictitious identity). Purchased in 1962.

Curtiss Kittyhawk I, Curtiss-Wright Corporation, U.S., 1942. Construction number 18780; registration AL135 (RAF), 1076 (RCAF). Transferred from the RCAF in 1964.

Curtiss Seagull, Curtiss Aeroplane and Motor Company, U.S., c. 1920. Construction number unknown; no registration. Acquired in 1968 through exchange.

Czerwinski/Shenstone Harbinger (sailplane), homebuilt by various constructors, 1949–75. Construction number C1; registration C-FZCS (Can.). Acquired as a donation in 1978 (Mr. A.N. Le Cheminant, Manotick, Ontario, and Mr. R. Noonan, Parry Sound, Ontario).

de Havilland Canada DHC-1B2 Chipmunk 2, de Havilland Aircraft of Canada Ltd., Canada, 1956. Construction number 208-246; registration 18070 (RCAF), 12070 (CAF), CF-CIA (Can.). Purchased in 1972.

de Havilland Canada DHC-2 Beaver, de Havilland Aircraft of Canada Ltd., Canada, 1947. Construction number 1; registration CF-FHB-X, CF-FHB, C-FFHB (Can.). Purchased in 1980.

de Havilland Canada DHC-3 Otter, de Havilland Aircraft of Canada Ltd., Canada, 1960. Construction number 370; registration 9408 (RCAF). Transferred from the Canadian Armed Forces in 1983.

de Havilland Canada DHC-6 Twin Otter Series 100, de Havilland Aircraft of Canada Ltd., Canada, 1965. Construction number 1; registration CF-DHC-X (Can.). Donated by de Havilland Canada Ltd. in 1981.

de Havilland Canada DHC-7 Dash 7, de Havilland Aircraft of Canada Ltd., Canada, 1975. Construction number 1; registration C-GNBX (Can.). Donated by de Havilland Canada Ltd. in 1988.

de Havilland DH.60X Cirrus Moth, de Havilland Aircraft Company Ltd., U.K., 1928. Construction number 630; registration G-CAUA (Can.). Acquired as a donation in 1962 (Mr. C.F. Burke, Charlottetown, P.E.I.).

de Havilland DH.80A Puss Moth, de Havilland Aircraft Company Ltd., U.K., 1931. Construction number 2187; registration HM534 (RAF), 8877 (U.S. Navy), G-AHLO (U.K.), CF-PEI (Can.). Purchased in 1976.

de Havilland DH.82C Tiger Moth (partial aircraft), de Havilland Aircraft of Canada Ltd., Canada, 1941. Construction number 724; registration 4394 (RCAF), CF-FGL (Can.). Purchased in 1962.

de Havilland DH.82C2 Menasco Moth, de Havilland Aircraft of Canada Ltd., Canada, 1941. Construction number 1052; registration 4861 (RCAF). Transferred from the RCAF in 1964.

de Havilland DH.83C Fox Moth, de Havilland Aircraft of Canada Ltd., Canada, 1947. Construction number FM-28/2; registration CF-DJB (Can.). Acquired as a donation in 1989 (Mr. Maxwell Ward, Edmonton, Alberta).

de Havilland DH.98 Mosquito B.MK.XX, de Havilland Aircraft of Canada Ltd., Canada, 1944. Construction number unknown; registration KB336 (RAF). Transferred from the RCAF in 1964.

de Havilland DH.100 Vampire I, English Electric Company Ltd., U.K., 1945. Construction number unknown; registration TG372 (RAF). Transferred from the Ontario Science Centre in 1968.

de Havilland DH.100 Vampire 3, English Electric Company Ltd., U.K., 1948. Construction number EEP42392; registration 17074 (RCAF). Transferred from the RCAF in 1964.

de Havilland DH.106 Comet (cockpit section), de Havilland Aircraft Ltd., U.K., 1953. Serial number 06017; registration 5301 (RCAF). Transferred from the RCAF in 1964.

Douglas DC-3, Douglas Aircraft Company Inc., U.S., 1942. Construction number 6261; registration 43-1985 (USAAF), CF-TDJ, C-FTDJ (Can.). Donated by Goodyear Corporation in 1983.

Easy Riser (powered hang-glider), Ultralight Flying Machines, U.S., 1977. Serial number and registration unknown. Acquired as a donation in 1995 (Mr. M. Pelletier, Orleans, Ontario).

Fairchild 82A, Fairchild Aircraft Ltd., Canada, 1937. Construction number 61; registration CF-AXL (Can.). Donated by Canadian Pacific Airlines in 1967.

Fairchild C-119F Flying Boxcar (cockpit section), Fairchild Engine and Airplane Corporation, U.S., date of manufacture unknown. Construction number and registration unknown. Transferred from the RCAF, date unknown.

Fairchild FC-2W-2, Fairchild Airplane Manufacturing Corporation, U.S., 1928. Construction number 128; registration NC6621 (U.S.), G-CART (Can.; fictitious identity). Donated by Aero Service Corporation in 1962.

Fairchild PT-26B Cornell III, Fleet Aircraft Ltd., Canada, 1942. Construction number FC 239; registration 10738 (RCAF). Transferred from the RCAF in 1964.

Fairey Battle IT, Fairey Aviation Co. Ltd., U.K., 1940. Construction number F-4848; registration R7384 (RAF). Transferred from the RCAF in 1964.

Fairey Firefly FR.1, General Aircraft Ltd., U.K., 1945. Construction number F.7776; registration DK545 (RN, RCAF, and RCN). Donated by the government of Eritrea in 1993.

Fairey Swordfish II, Blackburn Aircraft Co., U.K., c. 1943. Construction number unknown; registration NS122 (RCN; fictitious identity). Purchased by Canadian War Museum in 1965 and later transferred to the Canada Aviation Museum.

Fantasy 7 (balloon), Fantasy Sky Promotions Inc., Canada, 1981. Serial number 001; registration C-GEMW (Can.). Acquired as a donation in 1990 (Mr. J.R. Connacher, Toronto, Ontario).

Fleet 2/7 (parts of two aircraft), Fleet Aircraft Ltd., Canada, c. 1930. Fleet 2 components purchased in 1963. Construction numbers and registration numbers unknown for both aircraft.

Fleet 16B Finch II, Fleet Aircraft Ltd., Canada, 1940. Construction number 408; registration 4510 (RCAF), NC1327V (U.S.). Transferred from the RCAF in 1966.

Fleet 50K Freighter (parts of two aircraft), Fleet Aircraft Ltd., Canada, 1939. Serial number 202; registration CF-BJU (Can.), 799 (RCAF), CF-BXP (Can.). Donated by Labrador Mining and Exploration Ltd. and salvaged in 1964. Smaller parts of serial number 203, registration CF-BJW (Can.) salvaged and aquired by donation in 1970 (Mr. M.L. McIntyre, Toronto, Ontario).

Fleet 80 Canuck, Fleet Aircraft Ltd., Canada, 1946. Construction number 149; registration CF-EBE (Can.). Purchased in 1974.

Fokker D.VII, Fokker Flugzeugwerke GmbH, Germany, 1918. Construction number 3659; registration 10347/18 (German Air Service), 1178 (U.S.). Purchased in 1971.

Fokker Universal (engine and fuselage), Atlantic Aircraft Corporation, U.S., 1928. Serial number 434; registration NC7029 (U.S.). Acquired in 1998 through exchange.

Found FBA-2C, Found Brothers Aviation Ltd., Canada, 1963. Construction number 4; registration CF-OZV (Can.), C-GCCF (Can.; fictitious identity). Donated by Centennial College, Toronto, Ontario, in 1979.

Grumman CP-121 Tracker, de Havilland Canada Ltd., Canada, 1960. Construction number DHC-86; registration 1587 (RCN), 12187 (CAF). Transferred from the Canadian Forces in 1990.

Grumman Goose II, Grumman Aircraft Engineering, 1944. Construction number B-77; registration 37824 (USN), 391 (RCAF), CF-MPG (Can.), C-FMPG (Can.). Donated by the RCMP and the government of British Columbia in 1995.

Hawker Hind, Hawker Aircraft Ltd., U.K., 1937. Construction number unknown; registration L7180 (RAF). Donated by the government of Afghanistan in 1975.

Hawker Hurricane XII, Canadian Car and Foundry Company Ltd., Canada, 1942. Construction number 522/19308; registration 5584 (RCAF). Transferred from the RCAF in 1964.

Hawker Sea Fury FB.11, Hawker Aircraft Ltd., U.K., 1948. Construction number unknown; registration TG119 (RN and RCN). Donated by Bancroft Industries in 1963.

Hawker Siddeley AV-8A Harrier, Hawker Siddeley Aviation, U.K., 1973. Construction number 712127/49; registration 15866 (USMC). On loan from U.S. Marine Corps since 1997.

Heinkel He 162A-1 Volksjäger, Ernst Heinkel Flugzeugwerke GmbH, Germany, 1945. Construction number unknown; registration 120076 (Luftwaffe), VH523 (RAF), Air Ministry #59 (U.K.). Transferred from the RCAF in 1964 (fictitious unit markings).

Heinkel He 162A-1 Volksjäger, Ernst Heinkel Flugzeugwerke GmbH, Germany, 1945. Construction number unknown; registration 120086 (Luftwaffe), Air Ministry #62 (U.K.). Transferred from the RCAF in 1964.

Hispano HA-1112-M1L Buchón, La Hispano Aviación S.A., Spain, c. 1955. Construction number 164; registration C.4K-114 (Spanish Air Force). Purchased by the Canadian War Museum in 1967 and transferred to the Canada Aviation Museum in 1968.

Junkers J.I., Junkers-Fokker Flugzeugwerke A.G., Germany, 1918. Construction number 252; registration 586/18 (German Air Service). Transferred from the Canadian War Museum in 1969.

Junkers W.34f/fi, Junkers Flugzeugwerke A.G., Germany, 1931. Construction number 2718; registration CF-ATF (Can.). Acquired as a donation in 1962 (Mrs. J.A. Richardson, Winnipeg, Manitoba).

Lockheed 10A Electra, Lockheed Aircraft Corporation, U.S., 1937. Construction number 1112; registration CF-TCA (Can.). Donated by Air Canada in 1968.

Lockheed 12A Electra Junior, Lockheed Aircraft Corporation, U.S., 1937. Construction number 1219; registration CF-CCT (Can.). Donated by the Department of Transport in 1963.

Lockheed 1329 Jetstar 6, Lockheed Aircraft Corporation, U.S., 1961. Construction number 5018; registration N9290R (U.S.), CF-DTX, C-FDTX (Can.). Donated by Transport Canada in 1986.

Lockheed F-104A Starfighter, Lockheed Aircraft Corporation, U.S., 1957. Construction number 183-1058; registration 56-770 (USAF), 12700 (RCAF). Transferred from the Canadian Armed Forces in 1968.

Maurice Farman Série 11 Shorthorn, Aircraft Manufacturing Company, U.K., c. 1916. Construction number unknown; registration G-AUBC (Australia), VH-UBC (Australia), N9645Z (U.S.). Purchased in 1981.

McDonnell CF-101B Voodoo, McDonnell Aircraft Corporation, U.S., 1959. Construction number 518; registration 57-340 (USAF), 101025 (CAF). Transferred from the Canadian Armed Forces in 1984.

McDonnell F2H-3 Banshee, McDonnell Aircraft Corporation, U.S., 1953. Construction number 174; registration 126464 (USN and RCN). Donated by the Royal Canadian Navy in 1965.

McDonnell Douglas CF-188B (CF-18), McDonnell Douglas Aircraft Corporation, U.S., 1982. Construction number unknown; registration 188901 (CAF). Transferred from the Canadian Forces in 2001.

McDowall Monoplane, homebuilt by Robert McDowall, Canada, 1915. No construction number; no registration. Purchased in 1967.

Messerschmitt Bf 109F-4, Erla-Maschinenwerk, Germany, 1942. Construction number unknown; registration 10132 (Luftwaffe). Received through exchange in 1999.

Messerschmitt Me 163B-1a Komet, Klemm Technik GmbH, Germany, 1945. Construction number 315/1/1; registration 19114 (Luftwaffe), 191916 (Luftwaffe). Transferred from the RCAF in 1964.

Mikoyan Gurevich MiG-15bis (Lim-2), WSK Mielec, Poland, 1954. Construction number 1 B 003-16; registration 316 (Polish Air Force). Received through exchange in 1998.

Mitchell Wing B-10 (homebuilt), Mitchell Wing, Canada [1980s]. No construction number; no registration. Acquired as a donation in 2002 (Mr. M. Johnston, Nanaimo, British Columbia).

Moyes Stingray (hang-glider), Muller Kites, Canada, c. 1977. No construction number; no registration. Acquired as a donation in 2002 (Mr. M. Johnston, Nanaimo, British Columbia).

Nieuport 12, Société anonyme des établissements Nieuport, France, c. 1915. Construction number unknown; registration 1504 (Aéronautique militaire). Gift to Canada from the French government in 1917 and transferred from the RCAF in 1965.

Nieuport 17 (reproduction), Carl Swanson, U.S., 1963, rebuilt by the Canada Aviation Museum following crash, 1990–95. No construction number; registration CF-DDK (Can.), B1566 (Royal Flying Corps; fictitious identity). Acquired as a donation in 1962. On long-term loan to the Canadian War Museum.

Noorduyn Norseman VI, Noorduyn Aircraft Ltd., Canada, 1943. Construction number 136; registration 787 (RCAF). Transferred from the RCAF in 1964.

North American Harvard II, North American Aviation, U.S., 1940. Construction number 66-2565; registration 2532 (RCAF). Transferred from the RCAF in 1964.

North American Harvard II, North American Aviation, U.S., 1941. Construction number 81-4107; registration 3840 (RCAF). Transferred from the RCAF in 1961. On long-term loan to Atlantic Canada Aviation Museum.

North American Harvard 4, Canadian Car and Foundry Co. Ltd., Canada, 1952. Construction number CCF4-178; registration 20387 (RCAF), CF-GBV (Can.). Transferred from the RCAF in 1964.

North American P-51D Mustang IV, North American Aviation Inc., U.S., 1945. Construction number 122-39806; registration 9298 (RCAF; fictitious colours and markings). Transferred from the RCAF in 1964.

North American TB-25L Mitchell 3PT, North American Aviation Inc., U.S., 1945. Construction number 108-47453; registration 5244 (RCAF; fictitious colours and markings). Transferred from the RCAF in 1964.

Northrop Delta Mk.II (wreckage), Canadian Vickers Ltd., Canada, 1937. Construction number CV-183; registration 673 (RCAF). Recovered by Canadian Armed Forces in 1969 and donated to Canada Aviation Museum.

Piasecki HUP-3, Piasecki Helicopter Corporation, U.S., 1954. Construction number unknown; registration 51-116623 (U.S. Army and RCN). Purchased by the Canadian War Museum in 1965 and later transferred to the Canada Aviation Museum.

Pitcairn-Cierva PCA-2 (autogyro), Pitcairn Aircraft Inc., U.S., 1931. Construction number 8; registration NR26 (U.S.), NC2624 (U.S.). Purchased in 1969.

Pitts Special S-2A (modified), Aerotek Inc., U.S., 1973. Serial number 2059; registration CF-AMR (Can.), C-FAMR (Can.). Acquired as a donation in 2002 (Mr. Dave Gillespie, Gorman Park, Saskatchewan).

Rosie O'Grady's Balloon of Peace, E. Yost, U.S., c. 1984. No construction number; registration N53NY (U.S.). Acquired as a donation in 1986.

Royal Aircraft Factory BE 2C, British and Colonial Aeroplane Co. Ltd., U.K., 1915. Construction number B & C 1042; registration 5878 (RFC). Acquired by the Canadian government with other war trophies in 1919.

Rutan Quickie, Quickie Aircraft Corporation (kit), U.S., homebuilt by J.D. Vos, Canada, 1984. No construction number; registration C-GGLC (Can.). Acquired as a donation in 1990 (Mr. J.D. Vos, Hamilton, Ontario).

Sheldrake Merrill *Spirit of Canada* (balloon), homebuilt by Stanley Sheldrake, Canada, 1967. No construction number; registration CF-VOZ. Acquired as a donation in 1999 (Mr. Stanley Sheldrake, Smithville, Ontario).

Sikorsky H-5 Dragonfly (S-51), Sikorsky Aircraft, U.S., 1947. Construction number 5118; registration 9601 (RCAF). Transferred from the RCAF in 1964.

Sikorsky HO4S-3 Horse (S-55), Sikorsky Aircraft, U.S., 1955. Construction number 55877; registration 55877 (RCN). Transferred from the Canadian Armed Forces in 1970.

Sikorsky R-4B, Sikorsky Aircraft, U.S., 1944. Construction number unknown; registration 43-46565 (USAAF). Acquired in 1983 through exchange.

Sopwith 2F.1 Camel, Hooper & Company Ltd., U.K., 1918. Construction number unknown; registration N8156 (RAF; fictitious colours and markings). Transferred from the Canadian War Museum in 1967.

Sopwith 7F.1 Snipe, Nieuport & General Aircraft Corporation Ltd., U.K., 1918. Construction number unknown; registration E6938 (RAF). Purchased by the Canadian War Museum in 1964 and later transferred to the Canada Aviation Museum.

Sopwith Pup (reproduction), George Neal, Canada, 1967. Construction number C552; registration CF-RFC (Can.), B2167 (RFC; fictitious identity). Purchased in 1973.

Sopwith Triplane (reproduction), Carl Swanson, U.S., 1966. No construction number; registration N5492 (RNAS; fictitious identity). Purchased in 1966.

Spad VII, Mann Egerton & Company, U.K., 1917. Construction number 103; registration B9913 (RFC). Purchased in 1965.

Spectrum Beaver RX550, Spectrum Aircraft Inc., Canada, 1986. Construction number unknown; registration C-IGOW (Can.). Donated by Invacare Canada/Hovis Medical Ltd. in 1987.

Stearman 4EM Junior Speedmail, Stearman Aircraft Co., U.S., 1930. Construction number 4021; registration NC784H (U.S.), CF-AMB (Can.; fictitious colours and markings). Acquired as a donation in 1970 (Mr. John Paterson, Thunder Bay, Ontario).

Stinson SR Reliant, Stinson Aircraft Co., U.S., 1933. Construction number 8717; registration NC13464 (U.S.), CF-HAW (Can.), C-FHAW (Can.). Purchased in 1983.

Stits SA-3A Playboy (homebuilt), Keith Hopkinson, Canada, 1955. Construction number 5501; registration CF-IGK, CF-RAD (Can.), C-FRAD (Can.). Purchased in 1978.

Supermarine Spitfire Mk.IIB, Supermarine Division, Vickers Armstrong Ltd., U.K., 1941. Construction number CBAF 711; registration P8332 (RAF), A166 and 166B (RCAF). Transferred from the RCAF in 1968. On long-term loan to the Canadian War Museum.

Supermarine Spitfire L.F. Mk.IX, Supermarine Division, Vickers Armstrong Ltd., U.K., 1944. Construction number CBAF IX 2161; registration NH188 (RAF), H-109, H-64 (Royal Netherlands Air Force), SM-39 (Force aérienne belge), OO-ARC (#1077) (Belgian civil), CF-NUS (Can.; fictitious colours and markings). Aquired as a donation in 1964 (Mr. John Paterson, Thunder Bay, Ontario).

Supermarine Spitfire Mk.XVIe, Supermarine Division, Vickers Armstrong Ltd., U.K., 1945. Construction number CBAF IX 4424; registration TE 214 (RAF), TE 353 (RAF), TE 353 (RCAF), TE 214 (RCAF; fictitious colours and markings). Transferred from the RCAF in 1966.

Taylor E-2 Cub, Taylor Aircraft Co. Inc., U.S., 1935. Construction number 289; registration NC15399 (U.S.), C-GCGE (Can.). Acquired in 1985 through exchange.

Taylorcraft BC-65, Taylorcraft Aviation Corporation, U.S., 1939. Serial number 1409; registration CF-BPR (Can.), C-FBPR (Can.). Acquired as a donation in 1999 (Mr. Harry Drover, Collingwood, Ontario).

Travel Air 2000, Travel Air Manufacturing Co., U.S., 1929. Construction number 720; registration C6281 (U.S.), CF-AFG (Can.). Purchased in 1968.

Vickers Viscount 757, Vickers Armstrong Aircraft Ltd., U.K., 1957. Construction number 270; registration CF-THI (Can.). Donated by Air Canada in 1969.

Waco 10 (GXE), Advance Aircraft Co., U.S., 1928. Serial number 1521; U.S. registration unknown, registration C-GAFD (Can.). Donated by Leavens Aviation Inc. in 2002.

Westland Lysander III, Westland Aircraft Ltd., U.K., rebuilt in 1967 by the RCAF using three aircraft. Registration R9003 (RAF; fictitious identity). Donated by the Canadian Armed Forces in 1968.

Wills Wing XC-185 (hang-glider), Wills Wing Inc., U.S., 1977. No serial number. No registration. Acquired as a donation in 1982 (Mr. S. Midwinter, Calgary, Alberta).

Zenair CH-300 Tri-Zenith, Zenair Canada Ltd. (kit), home-built by R. Morris, G. Boudreau, and D. Holtby, Canada, 1978. Construction number 300; registration C-GOVK (Can.). Purchased in 1984.

Bibliography

Avery, Norman. *Whiskey Whiskey Papa: Chronicling the Exciting Life and Times of a Pilot's Pilot.* Ottawa: self-published, 1998.

Barris, Ted. *Behind the Glory.* Toronto: Macmillan, 1992.

Bashow, David L. *Starfighter: A Loving Retrospective of the CF-104 Era in Canadian Fighter Aviation, 1961–1968.* Stoney Creek: Fortress Publications, 1990.

Blatherwick, Francis John. *A History of Airlines in Canada.* Toronto: Unitrade Press, 1989.

Braun, Don C. with John C. Warren. *The Arctic Fox: Bush Pilot of the North Country.* Excelsior: Back Bay Press, 1994.

Bungey, Lloyd M. *Pioneering Aviation in the West: As Told by the Pioneers.* Surrey: Hancock House, 1992.

Campagna, Palmiro. *Storms of Controversy: The Secret Avro Arrow Files Revealed.* Toronto: Stoddart, 1992.

Chajkowsky, William E. *Royal Flying Corps: Borden to Texas to Beamsville.* Cheltenham, Ontario: Boston Mills Press, 1979.

Charlton, Peter, Leo Pettipas and Michael Whitby, eds. *"Certified Serviceable": Swordfish to Sea King; The Technical Story of Canadian Naval Aviation by Those Who Made It So.* Gloucester, Ontario: CNATH Book Project, 1995.

Christie, Carl. *Ocean Bridge: The History of RAF Ferry Command.* Toronto: University of Toronto Press, 1995.

Collins, David H. *Wings Across Time: The Story of Air Canada.* Toronto: Griffin House, 1986.

Conrad, Peter C. *Training for Victory: The British Commonwealth Air Training Plan in the West.* Saskatoon: Western Producer Prairie Books, 1989.

Corley-Smith, Peter. *Barnstorming to Bush Flying: British Columbia's Aviation Pioneers, 1910–1930.* Victoria: Sono Nis Press, 1989.

Corley-Smith, Peter. *Bush Flying to Blind Flying: British Columbia's Aviation Pioneers, 1930–1940.* Victoria: Sono Nis Press, 1993.

Corley-Smith, Peter and David N. Parker. *Helicopters: The British Columbia Story.* Toronto: Canav Books, 1985.

Dempsey, Daniel V. *A Tradition of Excellence: Canada's Airshow Team Heritage.* Victoria: High Flight Enterprises, 2002.

Dillon, J.C. *Early Days: A Record of the Early Days of the Provincial Air Service of Ontario, of the Men and the Ships They Flew.* Toronto: Department of Lands and Forests, 1961.

Dodds, Ronald. *The Brave Young Wings.* Stittsville, Ontario: Canada's Wings, 1980.

Douglas, W.A.B. *The Creation of a National Air Force.* Toronto: University of Toronto Press, 1986 (vol. 2 of *The Official History of the Royal Canadian Air Force*).

Douglas, W.A.B., translated by Jean Pariseau. *La création d'une aviation militaire nationale.* Ottawa: Ministère de la Défense nationale, 1987 (vol. 2 of *Histoire officielle de l'Aviation royale du Canada*).

Duffy, Dennis and Carol Crane. *The Magnificent Distances: Early Aviation in British Columbia, 1910–1940.* Victoria: Provincial Archives, 1980.

Ellis, Frank H. *Canada's Flying Heritage.* Toronto: University of Toronto Press, 1954.

Ellis, Frank H. *In Canadian Skies: 50 Years of Adventure and Progress.* Toronto: Ryerson Press, 1959.

Ellis, John R. *Canadian Civil Aircraft Register.* Toronto: Canadian Aviation Historical Society, 1975.

English, Allan D. *The Cream of the Crop: Canadian Aircrew, 1939–1945.* Montréal and Kingston: McGill-Queen's University Press, 1996.

Fischer Von Poturzyn, F.A., translated by Edward Morley. *Junkers and World Aviation: A Contribution to German Aeronautical History, 1909–1934.* Munich: R. Pflaum, 1935.

Fletcher, David C. and Doug MacPhail. *Harvard! The North American Trainers in Canada.* San Josef: DCF Flying Books, 1990.

Floyd, Jim. *The Avro Canada C102 Jetliner.* Erin: Boston Mills Press, 1986.

Foster, J.A. *The Bush Pilots: A Pictorial History of a Canadian Phenomenon.* Toronto: McClelland & Stewart, 1990.

Fuller, G.A., J.A. Griffin and K.M. Molson. *125 Years of Canadian Aeronautics: A Chronology, 1840–1965.* Willowdale: Canadian Aviation Historical Society, 1983.

Gibbs-Smith, Charles H. *Aviation: An Historical Survey from Its Origins to the End of World War II.* London: Her Majesty's Stationery Office, 1970.

Gilbert, Walter E. and Kathleen Shackleton. *Arctic Pilot: Life and Work on North Canadian Air Routes.* Toronto: Thomas Nelson, 1939.

Gordon, Stanley. *The History of Aviation in Alberta to 1955.* Wetaskiwin: Reynolds-Alberta Museum, 1985.

Gowans, Bruce W. *Wings over Calgary, 1906–1940.* Calgary: Historical Society of Alberta, 1990.

Grant, R.G. *Flight: 100 Years of Aviation.* New York: Dorling Kindersley, 2002.

Greenaway, Keith R. and Sidney E. Colthorpe. *An Aerial Reconnaissance of Arctic North America.* Ottawa: Joint Intelligence Bureau, 1948.

Greenaway, Keith R. and Moira Dunbar. *Arctic Canada from the Air.* Ottawa: Canada Defence Research Board, 1956.

Greenhous, Brereton. *The Crucible of War, 1939–1945.* Toronto: University of Toronto Press, 1994 (vol. 3 of *The Official History of the Royal Canadian Air Force*).

Greenhous, Brereton and Hugh A. Halliday. *Canada's Air Forces, 1914–1999.* Montréal: Art Global, 1999.

Griffin, John and S. Kostenuk. *RCAF Squadron Histories and Aircraft, 1924–1968.* Toronto: Hakkert, 1977.

Gunston, Bill, ed. *Chronicle of Aviation.* London: Chronicle Communications, 1992.

Gunston, Bill. *The Osprey Encyclopedia of Russian Aircraft* (2nd ed.). Oxford: Osprey, 1995.

Halliday, Hugh A. *Not in the Face of the Enemy: Canadians Awarded the Air Force Cross and Air Force Medal, 1918–1966.* Toronto: Robin Brass Studio, 2000.

Halliday, Hugh A. and Larry Milberry. *The Royal Canadian Air Force at War, 1939–1945*. Toronto: Canav Books, 1990.

Hatch, F.J. *Aerodrome of Democracy: Canada and the British Commonwealth Air Training Plan, 1939–1945*. Ottawa: Canadian Government Publishing Centre, 1983.

Hitchins, F.H. *Air Board, Canadian Air Force and Royal Canadian Air Force*. Ottawa: Canadian War Museum, 1972.

Hotson, Fred W. *De Havilland in Canada*. Toronto: Canav Books, 1999.

Jenkins, Dennis R. *F/A-18 Hornet: A Navy Success Story*. New York: McGraw-Hill, 2000.

Kealy, J.D.F. and E.C. Russell. *A History of Canadian Naval Aviation, 1918–1962*. Ottawa: Queen's Printer, 1965.

Leigh, Z. Lewis. *And I Shall Fly*. Toronto: Canav Books, 1985.

McGrath, Gerald and Louis Sebert, eds. *Mapping a Northern Land: The Survey of Canada, 1947–1994*. Montréal and Kingston: McGill-Queen's University Press, 1999.

MacKenzie, David. *Canada and International Civil Aviation, 1932–1948*. Toronto: University of Toronto Press, 1989.

Main, J.R.K. *Voyageurs of the Air: A History of Civil Aviation in Canada, 1857–1967*. Ottawa: Queen's Printer, 1967.

Milberry, Larry. *Air Transport in Canada*, 2 vols. Toronto: Canav Books, 1997.

Milberry, Larry. *Aviation in Canada*. Toronto: McGraw-Hill Ryerson, 1979.

Milberry, Larry. *The Avro CF-100*. Toronto: Canav Books, 1981.

Milberry, Larry. *The Canadair North Star*. Toronto: Canav Books, 1982.

Milberry, Larry. *The Canadair Sabre*. Toronto: Canav Books, 1986.

Milberry, Larry. *Canada's Air Force at War and Peace*. 3 vols. Toronto: Canav Books, 1999–2001.

Milberry, Larry, ed. *Sixty Years: The RCAF and CF Air Command, 1924–1984*. Toronto: Canav Books, 1984.

Mills, Stephen E. and James W. Phillips. *Sourdough Sky: A Pictorial History of Flight and Flyers in the Bush Country*. Seattle: Superior Publishing, 1969.

Mitchell, Kent A. *Fairchild Aircraft, 1926–1987*. Santa Ana, California: Narkiewicz/Thompson, 1997.

Molson, Kenneth M. *Canada's National Aviation Museum: Its History and Collections*. Ottawa: National Aviation Museum, 1988.

Molson, Kenneth M. *Pioneering in Canadian Air Transport*. Winnipeg: James Richardson, 1974.

Molson, Kenneth M. and Alfred J. Shortt. *The Curtiss HS Flying Boats*. Ottawa: National Aviation Museum, 1995.

Molson, Kenneth M. and H.A. Taylor. *Canadian Aircraft since 1909*. Stittsville, Ontario: Canada's Wings, 1982.

Morrison, Colin A. *Voyage into the Unknown: The Search and Recovery of Cosmos 954*. Stittsville, Ontario: Canada's Wings, 1983.

Munson, Kenneth. *Airliners between the Wars, 1919–1939.* London, U.K.: Blandford Press, 1972.

Myers, Patricia A. *Sky Riders: An Illustrated History of Aviation in Alberta, 1906–1945.* Saskatoon: Fifth House, 1995.

Oswald, Mary E. *They Led the Way: Members of Canada's Aviation Hall of Fame.* Wetaskiwin, Alberta: Canada's Aviation Hall of Fame, 1999.

Para Rescue Association of Canada. *That Others May Live: 50 Years of Para Rescue in Canada.* Astra: Air Transport Group, 1994.

Parkin, J.H. *Aeronautical Research in Canada, 1917–1957: Memoirs of J.H. Parkin.* Ottawa: National Research Council, 1983.

Parley, Kay. *Flash-in-the-Sky Boy: From the Letters, Manuscripts, and Published Works of William Wallace Gibson.* Saskatoon: self-published, 1967.

Pickler, Ron and Larry Milberry. *Canadair: The First 50 Years.* Toronto: Canav Books, 1995.

Pigott, Peter. *Flying Canucks: Famous Canadian Aviators.* Toronto: Houslow, 1994.

Pigott, Peter. *National Treasure: The History of Trans Canada Airlines.* Madeira Park: Harbour Publishing, 2001.

Pigott, Peter. *Wingwalkers: A History of Canadian Airlines International.* Madeira Park: Harbour Publishing, 1998.

Ralph, Wayne. *Barker, VC: William Barker, Canada's Most Decorated War Hero.* Toronto: Doubleday Canada, 1997.

Render, Shirley. *No Place for a Lady: The Story of Canadian Women Pilots, 1928–1992.* Winnipeg: Portage and Main Press, 1992.

Roberts, Leslie. *There Shall Be Wings: A History of the Royal Canadian Air Force.* Toronto: Clarke, Irwin, 1959.

Rossiter, Sean. *The Immortal Beaver: The World's Greatest Bush Plane.* Vancouver: Douglas & McIntyre, 1996.

Rossiter, Sean. *Otter and Twin Otter: The Universal Airplanes.* Vancouver: Douglas & McIntyre, 1999.

Scott, Phil. *The Shoulders of Giants: A History of Human Flight to 1919.* Don Mills: Addison-Wesley, 1995.

Shaw, S. Bernard. *Photographing Canada from Flying Canoes.* Burnstown: General Store Publishing, 2001.

Simmons, George. *Target: Arctic—Men in the Skies at the Top of the World.* New York: Chilton Books, 1965.

Stevenson, Garth. *The Politics of Canada's Airlines from Diefenbaker to Mulroney.* Toronto: University of Toronto Press, 1987.

Sullivan, Alan. *Aviation in Canada, 1917–1918: Being a Brief Account of the Work of the Royal Air Force, Canada, the Aviation Department of the Imperial Munitions Board, and the Canadian Aeroplanes Limited.* Toronto: Rous and Mann, 1919.

Sullivan, Kenneth H. and Larry Milberry. *Power: The Pratt & Whitney Canada Story.* Toronto: Canav Books, 1989.

Sutherland, Alice Gibson. *Canada's Aviation Pioneers: 50 Years of McKee Trophy Winners.* Toronto: McGraw-Hill Ryerson, 1978.

Taylor, Michael J.H. *The Aerospace Chronology.* London: Tri-Service Press, 1989.

Thomson, Don W. *Men and Meridians: The History of Surveying and Mapping in Canada,* 3 vols. Ottawa: Queen's Printer, 1966–69.

Thomson, Don W. *Skyview Canada: A Story of Aerial Photography in Canada.* Ottawa: Energy, Mines and Resources Canada, 1975.

Turner, P. St. John and Heinz J. Nowarra. *Junkers: An Aircraft Album.* London: Ian Allan, 1971.

Ward, Max. *The Max Ward Story: A Bush Pilot in the Bureaucratic Jungle.* Toronto: McClelland & Stewart, 1991.

Waters, Andrew W. *All the U.S. Air Force Airplanes, 1907–1983.* New York: Hippocrene Books, 1983.

West, Bruce. *The Firebirds: An Account of the First 50 Years of the Ontario Provincial Air Service.* Toronto: Ministry of Natural Resources, 1974.

White, Howard and Jim Spilsbury. *The Accidental Airline: Spilsbury's QCA.* Madeira Park: Harbour Publishing, 1988.

Wise, S.F. *Canadian Airmen and the First World War.* Toronto: University of Toronto Press, 1980 (vol. 1 of *The Official History of the Royal Canadian Air Force*).

Photo Credits

Canada Aviation Museum: pages viii (Malak), 6 (21635), 7, 8 (Helene Croft, AL0130), 9 (top: 13420, bottom: 22691), 11, 15 (Austro Hungarian Collection), 16 (Austro Hungarian Collection), 19 (2177), 21 (1346), 22 (Robert Bradford, 1967.0893), 23 (31822), 24 (8752), 27 (6932), 30 (James Leech, 1990.0469.001), 33 (1389), 35 (4210), 38 (top: 1835, bottom: 19800), 39 (John Matthews), 43 (Robert Bradford, 1967.0884), 51 (4734), 52 (1491), 54 (4711), 55 (1158), 56 (Robert Bradford, 1967.0892), 59 (4691), 60 (26005), 61 (top: 5241, bottom: 4999), 65 (1238), 66 (Robert Bradford, 1967.0899), 67 (12527), 69 (6757), 70 (Robert Bradford, 24079), 73 (24811), 77 (bottom: 16220), 80 (John Matthews), 84 (John Matthews), 94 (5097), 95 (25771), 98 (30492), 100 (31698), 101 (28481), 103 (top: Raphael Beaussart, bottom: 28636), 106 (11540), 107 (11043B), 108 (bottom: 14537), 111 (24953), 112 (28648), 115 (John Davies), 130 (John Matthews), 135 (Ray Cavan), 138 (27598), 139 (top: 4598, bottom: 11457), 142 (5582), 146 (top left and bottom right: Molson Collection), 149 (Frank Oord, 1994.0550.001), 150 (23658), 151 (top: 11632, bottom: 13401), 153 (John Matthews), 154 (Air Canada Collection), 155 (Air Canada Collection), 156, 157 (23636), 160 (8343), 161 (Robert Bradford, 1967.0894), 164 (3091), 166 (top and bottom: John Matthews), 169 (John Matthews), 171 (31420), 172 (15738), 173 (21894), 174 (Robert Bradford, 1967.0882), 176 (John Matthews), 177 (18430), 178 (4551), 179 (10079), 180 (23472), 181 (Robert Bradford, 1967.0891), 182 (John Matthews), 183, 184 (John Matthews), 185 (top and centre: Raphael Beaussart; bottom: John Matthews), 190 (15050), 192 (Robert Bradford, 1967.0881), 194 (33052), 195 (2828), 197 (2442), 198 (2943), 199 (Robert Bradford, 1967.0897), 200 (5940), 201 (5931), 203 (4947), 207 (11696), 208 (Charles Vinh, 1992.2307), 209 (John Matthews), 210 (3732), 211 (35-021/10), 212 (4077), 215 (28477), 216 (32544), 217 (32543)

Canada Aviation Museum Library: pages 47, 123, 146 (top right and bottom left), 162 (both images), 170

Air Canada: page 145 (Robert Bradford)

Bombardier Inc.: page 218

Bombardier Regional Aircraft: page 219

Canada's Aviation Hall of Fame: page 48 (Robert Bradford)

© Canadian War Museum: page 44 (19780702-089), 45 (Robert Bradford, Beaverbrook Collection of Art), 82 (Paul Goranson, Beaverbrook Collection of Art, AN 19710261-3763), 86 (Robert Bradford, Beaverbrook Collection of Art, 197), 204 (left: AN 19910238-880, right: AN 19910238-881), 205 (AN 19920166-003)

Cougar Helicopters: page 116

Department of National Defence: pages 34, 41 (PL 23609), 46 (0-1744 AH470 "B"), 77 (top: RE-13082), 81 (PL-44203-A), 83 (PL 19507-4), 85 (PL 12377), 88 (PL 12754), 89 (top: PL 5952, bottom: PL 25471), 91 (top: PL 12062, bottom: PL 12764), 104 (HSC 92-0849-221), 108 (top: PCN 3219), 110 (ISC 83-536), 114 (TN70-1251), 118 (PCN 3915), 120 (RE-1655-3), 121 (PL 65305), 122 (PL 83510), 125 (PCN84-17), 126 (AEC83-1213), 127 (PL 101496), 128 (PCN 734), 129 (PL 51375), 133 (PL 131413), 186 (5356)

© Fred W. Hotson: page 202 (Robert Bradford)

Kirkman family: page 29

Leavens family: page 163

Library and Archives Canada: pages 3 (top: PA-59985, bottom: PA-062293), 4 (PA-62895), 72 (PA 63838), 99 (PA 133296), 148 (left: C-061893; top right: C-61899), 152 (C-063285), 187 (C-65073)

© David Metcalfe: page 57 (Robert Bradford)

Dan Patterson: pages 12, 20 (both photos), 26, 37, 40, 50, 53 (both photos), 58, 62, 74, 75, 78, 92, 96, 132, 136, 140, 141, 143, 158, 188, 196

Toronto Public Library: page 165

Index

Italic page numbers indicate information in pictures or captions.

629.1309710904 Can
Canadian wings.
PRICE: $60.00 (4141/)